SONET / SDH DEMYSTIFIED

Mc GRAW-HILL
TELECOMMUNICATIONS

Sonet / SDH Demystified

Steven Shepard

McGraw-Hill
New York   Chicago   San Francisco
Lisbon   London   Madrid   Mexico City
Milan   New Delhi   San Juan   Seoul
Singapore   Sydney   Toronto

McGraw-Hill

A Division of The McGraw-Hill Companies

ISBN 0-07-137618-6

*The sponsoring editor for this book was Stephen S. Chapman and the production
supervisor was Pamela A. Pelton. It was set in Century Schoolbook by MacAllister
Publishing Services, LLC.*

Printed and bound by Quebecor/Martinsburg.

This book is printed on recycled, acid-free paper containing a minimum of 50
percent recycled de-inked fiber.

McGraw-Hill books are available at special quantity discounts to use as premiums
and sales promotions, or for use in corporate training programs. For more information,
please write to the Director of Special Sales, McGraw-Hill, Two Penn Plaza,
New York, NY 10121-2298. Or contact your local bookstore.

Mom, thanks for teaching me strength, kindness, wisdom, and humor in this life. You've given me so much; someday I hope people will compare me to you. That would be the greatest honor.

CONTENTS

Contents

Contents

PREFACE

When I first contemplated the task of writing this book on SONET and SDH, my initial halting response was based on the fact that several wonderfully complete books were already written on those technologies by such accomplished writers as Walter Goralski and Stamatios Kartalopoulos. The more I thought about the project, however, the more I realized that there was room for another work that didn't focus so much on the technology as it did on the evolving global public network architecture and the equally evolving roles of SONET, SDH, and other optical networking technologies, as well as on the services and value derived from them—in short, the marketplace that the technologies served.

The telecommunications industry is currently trapped in the grip of a powerful sea change that promises to redefine the role of all players—manufacturers and service providers alike. In recent years (I could almost say months!), we have seen a full-circle turnabout as corporate focus on infrastructure design has gone from private line to ring architectures to mesh transport—in many ways a reinvention of private line transport. In each case, SONET and SDH play a central role. The reason for this ongoing evolution is the simultaneous reinvention of the services transported across the network and the changes in the roles of the customers who create those services.

A relatively small collection of forces has joined their considerable influence to drive this evolution to an as yet unknown conclusion. The first of these is technology convergence, which is the inexorable movement toward a single-protocol transport fabric, probably IP; the second, the currently unslakeable demand for broadband access and transport; the third, a strong demand for higher-bandwidth wireless access; and finally, the universal recognition by customers and service providers alike that the key to success is a focus on services and the intelligent deployment of technology to create customer value, rather than on the technology itself.

Technology convergence is one-third of the convergence troika that I described in an earlier book, *Telecommunications Convergence.* The troika comprises technology convergence, which currently seems to conclude with the deployment of an all-IP network riding on top of an ATM switching infrastructure and SONET/SDH transport; *company convergence,* characterized by the feeding frenzy that has been underway as companies frenetically buy each other as a way to create converged capability sets; and *services convergence,* the ultimate goal of delivering a complete set of converged services that absolutely satisfies the demands of customers. I like the following model of the cloud shown to illustrate this phenomenon.

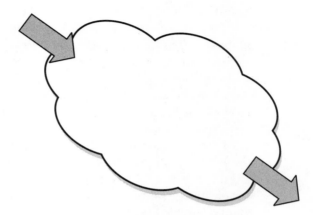

The upper left-hand arrow is the *G'Zinta*, whereas the lower right-hand arrow is the *G'Zouta*. The cloud model illustrates convergence at its best. If a customer transmits quality-sensitive traffic into the service provider cloud (the G'Zinta), it will arrive, unaltered, at the other end (the G'Zouta). In other words, the message that a service provider now conveys to the customer is "Look, for years we have operated in a world where you, the customer, have felt compelled to understand the underlying technologies in our networks. Well, today we change that. From now on, *you* concentrate on your business processes and let *us* worry about the network. That way we both do what we do best. Furthermore, as the service provider, I make the following commitment: I will endeavor to better understand the nature of your business so that I can offer access and transport services that help you meet the challenges of your *own* customers."

So what does all this have to do with SONET and SDH? A lot, actually. As demand for broadband access and transport grows and as the need for *quality of service* (QoS) increases, transport infrastructures must change to meet those demands. In the early 1980s, SONET and SDH were introduced as ways to satisfy the nascent demand for high-speed service that was just starting to show up on the distant horizon. Of course, it was clearly understood that they were *central office-to central office* (CO-to-CO) trunking technologies; customers would never have a need to even be aware of their existence because after all, what customer could *ever* have a need for that kind of bandwidth—I mean, come on!

The rest, as they say, is history, and how silly that history has been. As soon as customers became aware that the bandwidth was available, they naturally developed an almost immediate need for it. Soon thereafter, as convergence began to occur, they came up with an equally pressing need for real and measurable quality of service. SONET and SDH provided a large part of that, equipped with well-developed overhead designed to provide network control and rudimentary QoS over largely dedicated networks.

Today, however, things have changed. Many believe that SONET and SDH are far too overhead-heavy and inflexible to be part of the future network architecture; others argue that because they have been part of the network mix for so long, they deserve an incumbent's seat at the technology pantheon.

SONET and SDH are deeply embedded in the network's anatomy and will not disappear any time soon. A common mantra heard throughout the telephony world is, "If it ain't broke, don't fix it." Well, today that might well be changed to read, "If it ain't broke, look again. It might need a little tweaking."

This book is about SONET and SDH—no doubt about it. To focus strictly on how they function, however, would be to create an incomplete work, given the renewed focus on how revenue-generating network services can be derived from network technologies. To that end, we will cover the following major topics (among others):

- The evolution of bandwidth demand and the SONET and SDH technologies
- The evolving transport marketplace
- Supporting technologies, such as ATM and IP
- Evolving network architectures: point-to-point, ring, and mesh, and combinations of the three
- Service provider strategies for the deployment of broadband access and transport networks

Readers of my previous books may see some thematic similarities with this one. The themes that I developed in those earlier works still apply today, more, in fact, than ever before. The recent meltdown of the technology stocks pointedly illustrates the fact that technology, without an immediate, revenue-generating application in front of it, is not an ideal investment model. Today, the market's spotlight is brightly illuminating those firms that help technology companies create value behind their bit, byte, frame, and cell-oriented products. It is no wonder that, although they have certainly seen reduced revenues as part of the technology crash, companies like Siebel and others that have focused on enhancing the corporate value-chain have not been hit quite as hard as those focused purely on technology. Similarly, content providers and transport companies aligned with content providers have been somewhat shielded from the market's backlash because of their focus on tangible, recognizable, and realizable value.

Enough soapbox: Enjoy the book. As always, I look forward to your comments. Please send e-mail to **Steve@ShepardComm.com**.

STEVEN SHEPARD
Williston, Vermont

ACKNOWLEDGMENTS

Throughout the excitement of writing this book and the two titles that preceded it, I have been blessed with a collection of friends who in various ways helped immeasurably in the conversion of a concept to a book. I therefore thank the following people for their untiring efforts: Cyril Berg, Martha Bradley, Dave Brown, Rich Campbell, Joe Cappetta, Phil Cashia, Carmine Ciotola, Brian Clouse, Floyd Cross, Bob Dean, Mike Diffenderfer, Mark Fei, Jack Garrett, Jack Gerrish, Pathmal Gunawardana, Glenn Harrington, Dave Hill, Steve Hillier, Carol Hrobon, Barbara Jorge, Gary Kessler, Bill Kless, Naresh Lakhanpal, Johannes Lüthi, Gary Martin, John Hanschu, Mitch Moore, Richie Parlato, Todd Quam, Marta Ramirez, Mary Regan, Kenn Sato, Kirk Shamberger, Henry Sherwood, Elvia Szymanski, Christine Troianello, Ken Wade, John Wells, Sue Wetherell, and Dave Whitmore.

I am also lucky to have Steve Chapman in my corner, my fine and patient editor at McGraw-Hill and all around good guy. Thanks, Steve.

Finally, as always, to my family, for putting up with all the absences, even when I was there. You are the best and I love you very much.

Beginnings

The year was 1984, and Bill McGowan had a problem. Freshly bloodied and scarred from the AT&T divestiture battlefield, where as the head of MCI, he served as the attacking general who, in many people's minds, single-handedly drove the breakup of the Bell System. McGowan realized that the toppling of the titan and subsequent shattering of AT&T into eight distinct pieces (seven regional providers plus one long-distance provider), shown in Figure 1-1, only resolved *one* of the challenges that would lead to the creation of a truly competitive marketplace.

Although the best-known impact of divestiture was the breakup of AT&T (one result of which was the liberalization of the telecommunications marketplace in the U.S.), a second decision that was tightly intertwined with the Bell System's breakup was largely invisible to the public, yet was at least as important to AT&T competitors, MCI and Sprint, as the breakup itself. This decision, known as Equal Access, had one seminal goal: to make it possible for end customers to take advantage of one of the products of divestiture, the ability to select one's long-distance provider from a pool of available service providers—in this case AT&T, MCI, or Sprint. This, of course, was the realization of a truly competitive marketplace in the long-distance market segment.

To understand this evolution, it is helpful to have a high-level understanding of the overall architecture of the network. In the pre-divestiture world, AT&T was *the* provider for local service, long-distance service, and communications equipment. An AT&T *central office* (CO), therefore, was awash in AT&T hardware, such as switches, cross-connect devices, multiplexers, amplifiers, repeaters, and myriad other devices.

Figure 1-2 shows a typical network layout in the pre-divestiture world. A customer's telephone is connected to the service provider's network by a local loop connection (so-called twisted pair wire). The local loop, in turn, connects to the local switch in the central office. This switch is the point at

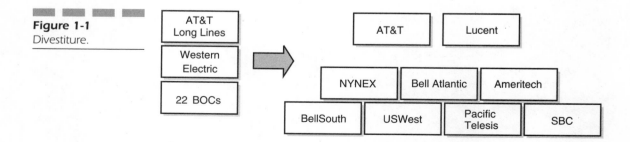

Figure 1-1
Divestiture.

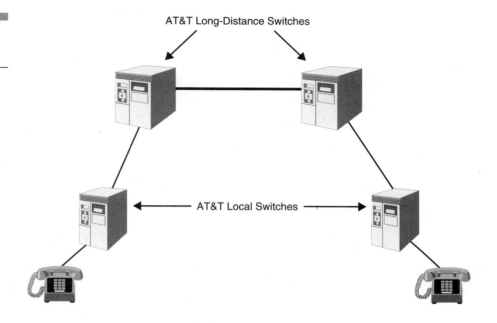

which customers first touch the telephone network, and it has the responsibility to perform the initial call setup, maintain the call while it is in progress, and tear it down when the call is complete. This switch is called a local switch because its primary responsibility is to set up local calls that originate and terminate within the same switch. It has one other responsibility, though, and that is to provide the necessary interface between the local switch and the long-distance switch, so that calls between adjacent local switches (or between far-flung local switches) can be established. The process goes something like this. When a customer lifts the handset and goes off-hook, a switch in the telephone closes, completing a circuit that enables current flow that in turn brings dial tone to the customer's ear. Upon hearing the dial tone, the customer enters the destination address of the call (otherwise known as a telephone number). The switch receives the telephone number and analyzes it, determining from the area code and prefix information whether the call can be completed within the local switch or must leave the local switch for another one. If the call is indeed local, it merely burrows through the crust of the switch and then reemerges at the receiving local loop. If the call is a toll or long-distance call, it must burrow through the hard crunchy coating of the switch, pass through the soft, chewy center, and emerge again on the other crunchy side on its way to a long-distance switch. Keep in mind that the local switch has no awareness of the existence of customers or telephony capability beyond its own cabinets. Thus, when it receives a telephone number that it is incapable of processing, it hands it off to a higher-order switch, with the implied message, "Here—I have no idea what to do with this, but I assume that you do."

The long-distance switch receives the number from the local switch, processes the call, establishes the necessary connection, and passes the call on to the remote long-distance switch over a long-distance circuit. The remote long-distance switch passes the call to the remote local switch, which rings the destination telephone and ultimately, the call is established.

Please note that in this pre-divestiture example, the originating local loop, local switch, long-distance switch, remote local loop, and all of the interconnect hardware and wiring belong to AT&T. They are all manufactured by Western Electric, based on a set of internal manufacturing standards that, if other manufacturers in the industry were there, would be considered proprietary. Because AT&T was the only game in town prior to divestiture, AT&T created the standard for transmission interfaces.

Fast forward now to January 1, 1984, and put yourself into the mind of Bill McGowan, whose company's survival depended upon the successful implementation of Equal Access. Unfortunately, Equal Access had one very serious flaw. Keep in mind that because the post-1984 network was emerging from the darkness of monopoly control, all of the equipment that comprised the network infrastructure was bought at the proverbial company store and was, by the way, proprietary.

Consider the newly re-created post-divestiture network model shown in Figure 1-3. At the local switch level, little has changed. At this point in time, only a single local services provider is available. At the long-distance level, however, a significant change has occurred. Instead of a single long-distance service provider called AT&T, three are now available: AT&T, MCI, and Sprint. The competitive mandate of Equal Access was designed to guarantee that a customer could freely select his or her long-distance provider of choice. If he or she wanted to use MCI's service instead of AT&T's, a simple call to the local telephone company's service representative would result in the generation of a service order that would cause the customer's local service to be logically disconnected from AT&T and reconnected to MCI. This way, long-distance calls placed by the subscriber would automatically be handed off to MCI. The problem of Equal Access to customers for the three long-distance providers was thus solved—almost.

Figure 1-3
The post-divestiture switching hierarchy and Equal Access.

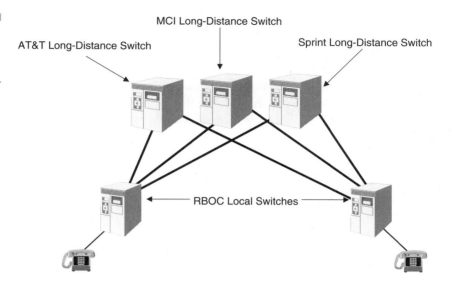

The proprietary nature of the network infrastructure that MCI and Sprint had to connect themselves to still existed and this was the root of McGowan's problem. The Equal Access amendment mandated that AT&T create space in its now RBOC-owned central offices for the interconnect equipment that MCI and Sprint required to establish *points-of-presence* (POPs) so that they could interconnect with the local switching equipment and therefore the customers. Unfortunately, because of the proprietary nature of the single-vendor world into which they were inserting themselves, both MCI and Sprint were required to buy AT&T's communications equipment in order to connect to the incumbent network infrastructure. Neither MCI nor Sprint were interested in pouring money into the pockets of AT&T, which is why McGowan had a problem. He did not want to be obligated to put jingle into AT&T's pockets simply to satisfy the interoperability requirement, so he took his case before a series of standards bodies, including the Interexchange Carrier Compatibility Forum, Bellcore (now Telcordia), ANSI, the CCITT (now the ITU-T), and a variety of other regional and international bodies. He argued his case effectively before them, claiming that requiring MCI and Sprint to purchase AT&T hardware was unfair. They agreed and tasked themselves to create a standard that would provide for true, open vendor interoperability. That standard, over the course of the ensuing eight years, became the *Synchronous Optical Network* (SONET) and ultimately, the *Synchronous Digital Hierarchy* (SDH). It is synchronous because the send and receive devices for the most part dance to the same network timing tune, and optical because the standards are designed to operate over fiber.

So what does this mean? At the time of divestiture, network providers faced the following challenges. No equipment standards existed, so mid-span meet, another term for interoperability, was virtually non-existent. Network management, a singularly important capability, was primitive, crude, and highly unreliable. Standards-based digital transmission for all intents and purposes ended at DS3. Networking, such as it was, was fragmented, a patchwork quilt of technologies and network fabrics that exchanged traffic by dint of brute force more than anything else. Standards for optical multiplexing, a key component of high-speed, long-haul networks, were emergent and proprietary.

With the arrival of SONET and SDH, most of those problems evaporated. Interoperability became a non-issue; seamless integration of legacy and emerging optical technology solutions became the rule rather than the exception; and network management's capabilities, once rudimentary and intermittent, expanded dramatically.

So with the arrival of SONET and SDH, networks took on an entire new set of behaviors that added to the capabilities of service providers and end customers alike. However, even in the pre-SONET/SDH environment, communications networks were remarkably capable. So before we dive into the details of SONET and SDH transmission, let's first go back and look at the networks that preceded them.

The Voice Network

The original voice network, including access, transmission facilities, and switching components, was exclusively analog until 1962, when T-Carrier emerged as an intra-office trunking scheme. The technology was originally introduced as a short-haul, four-wire facility to serve metropolitan areas. Over the years, it evolved to include coaxial cable facilities, digital microwave systems, fiber, and satellite.

As the network topology improved, so did the switching infrastructure. In 1976, AT&T introduced the 4ESS switch primarily for toll applications and followed it up with the 5ESS in 1981 for local switching access, as well as a variety of remote switching capabilities. Nortel, Siemens, and Ericsson all followed suit with equally capable hardware.

The goal of digitizing the human voice for transport across an all-digital network grew out of work performed at Bell Laboratories shortly after the turn of the century. That work led to a discrete understanding of not only the biological nature and spectral makeup of the human voice, but also to a better understanding of language, sound patterns, and the sounded emphases that comprise spoken language.

The Nature of Voice

A typical voice signal comprises frequencies that range from approximately 30 Hz to 10 KHz. Most of the speech energy, however, lies between 300 Hz and 3,300 Hz, the so-called voice band. Experiments have shown that the frequencies below 1 KHz provide the bulk of recognizability and intelligibility, whereas the higher frequencies provide richness, articulation, and natural sound to the transmitted signal.

The human voice comprises a remarkably rich mix of frequencies, but this richness comes at a considerable price. In order for telephone networks to transmit voice's entire spectrum of frequencies, significant network bandwidth must be made available to every ongoing conversation. A substantial price tag is attached to bandwidth; it is a finite commodity within the network, and the more of it that is consumed, the more it costs.

The Network

Thankfully, work performed at Bell Laboratories at the beginning of the 20th century helped network designers confront this challenge head-on. To understand it, let's take a tour of the telephone network.

The typical network, as shown in Figure 1-4, is divided into several regions: the access plant, the switching, multiplexing, and circuit connectivity equipment (the central office), and the long-distance transport plant. The access and transport domains are often referred to as the *outside plant*; the central office is, conversely, the *inside plant*. The outside plant has the responsibility to aggregate inbound traffic for switching and transport across the long-haul network, as well as to terminate traffic at a particular destination. The inside plant, on the other hand, has the responsibility to multiplex incoming traffic streams, switch the streams, and select an outbound path for ultimate delivery to the next central office in the chain or the final destination.

Let's examine each region of the network.

Outside Plant   The most common form of network access is via a single pair of twisted wire that connects the customer's telephone to the central office. The pairs of wire that run to each home or business are aggregated

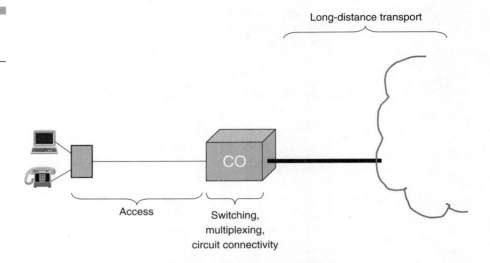

into a cable that runs from the neighborhood or downtown business district back to the central office that serves that area. The cables may be deployed as either underground plant, in which case, they are buried in conduit (see Figure 1-5), or aerial, in which case, they are suspended on telephone poles (see Figure 1-6).

Ultimately, the wire pairs, bundled as cables, make their way back to the central office where they are once again broken out and connected to the local switch so that they can be individually switched as required.

Inside Plant When the cables first enter the central office, they pass into a subterranean chamber in the basement of the CO called the *cable vault* (see Figure 1-7). At this point, the large cables are broken down into smaller cables (see Figure 1-8), after which they leave the cable vault and climb into the technological rafters of the office.

On a higher floor, the cables are dissected into their composite pairs, and each pair is then attached to electrical appearances on a large iron rack known as the *main distribution frame* (MDF) (see Figure 1-9). From the MDF, the pairs are interconnected to the switch, giving it the capability to establish demand connections from any pair in the office to any other pair in the office, as well as to a long-distance trunk should the need arise. This is shown in Figure 1-10.

Figure 1-5
Buried outside plant.

Figure 1-6
Aerial outside plant.

Figure 1-7
The cable vault.

Figure 1-8
Cable differentiation
in the cable vault.

Beginnings

Figure 1-9
Central office and
main distribution
frame.

Figure 1-10
Interconnection in
the modern central
office.

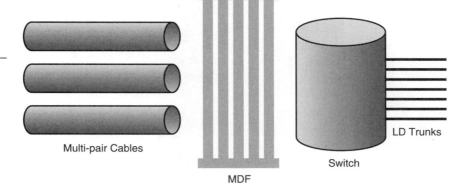

Multi-pair Cables

MDF

Switch

LD Trunks

The Early Years

Of course, networks didn't always work this way. When the telephone network first began to spread its tentacles across the continent, no switches were available. Initially, subscribers in towns bought individual phones and phone lines to each person they had a need to speak with, resulting in the famous image, shown in Figure 1-11, of metropolitan telephone poles with tier upon tier of cross-pieces, festooned with aerial wire, and its mathematical representation: n(n–1)/2, the total number of circuits that would be required for *n* people to be able to speak with everyone else in the community. For example, if 100 people in a neighborhood wanted to be able to speak with each other, they would require the phone company to install a total of (100)/(99)/2, or 4,950 circuits. Clearly, this was economically out of the question, not to mention the fact that the quantity of aerial cable threatened to block out the sun (see Figure 1-12), potentially precipitating the next Ice Age and the end of civilization as we know it.

The solution to this quandary came in two forms: central office switches and multiplexing. Central office switches enabled the phone company to provide the same level of connectivity, but required each customer to have

Figure 1-11
Early telephone pole showing large number of crossbars.

Figure 1-12
Mass of aerial cable.

only *one* circuit. Responsibility for setting up the connection to the called party resided with the switch in the central office, rather than with the customer. The first switches, of course, were operators, such as those shown in Figure 1-13; true mechanical switches didn't arrive until 1892, when Almon Strowger's Step-by-Step switch was first installed by Automatic Electric.

Strowger's story is worth telling because it illustrates the serendipity that characterized so much of this industry's development. It seems that Almon Strowger was not an inventor, nor was he a telephone person. He was an undertaker in a small town in Missouri. One day, he came to the realization that his business was (OK, we won't say dying) declining, and upon closer investigation, determined that the town's operator was married to his competitor! As a result, any calls that came in for the undertaker naturally went to her husband, not to Strowger.

To equalize the playing field, Strowger called upon his considerable talents as a tinkerer and designed a mechanical switch and dial telephone, shown in Figure 1-14, which are still in use today in a number of developing countries.

Figure 1-13
Operators at early switchboard.

Figure 1-14
A Strowger Telephone.

Multiplexing

Equally important as the development of the central office switch was the concept of multiplexing, which enabled multiple conversations to be carried simultaneously across a single shared physical circuit. The first such systems used *frequency-division multiplexing* (FDM), a technique made possible by the development of the vacuum tube, in which the range of available frequencies is divided into chunks that are then parceled out to subscribers. For example (and this is only an example), Figure 1-15 illustrates that subscriber #1 might be assigned the range of frequencies between 0 and 4,000 Hz, whereas subscriber #2 is assigned 4,000 to 8,000 Hz, #3 is assigned 8,000 to 12,000 Hz, and so on, up to the maximum range of frequencies available in the channelized system. In frequency-division multiplexing, we often observe that users are given "some of the frequency all of the time," meaning that they are free to use their assigned frequency allocation at any time, but may not step outside the bounds given to them. Early FDM systems were capable of transporting 24-4 KHz channels, for an overall system bandwidth of 96 KHz. Frequency-division multiplexing, although largely replaced today by more efficient systems that we will discuss later, is still used in analog cellular telephone and microwave systems, among others.

Figure 1-15
Frequency division multiplexing.

Subscriber 1: 0–4,000 Hz

Subscriber 2: 4,000–8,000 Hz

Subscriber 3: 8,000–12,000 Hz

Subscriber 4: 12,000–16,000 Hz

Subscriber 5: 16,000–20,000 Hz

This model worked well in early telephone systems. Because the lower regions of the 300–3,300 Hz voiceband carry the frequency components that provide recognizability and intelligibility, telephony engineers concluded that, although the higher frequencies enrich the transmitted voice, they are not necessary for calling parties to recognize and understand each other. This understanding of the makeup of the human voice helped them create a network that was capable of faithfully reproducing the sounds of a conversation while keeping the cost of consumed bandwidth to a minimum. Instead of assigning the full complement of 10 KHz to each end of a conversation, they employed filters to bandwidth-limit each user to approximately 4,000 Hz, a resource savings of some 60 percent. Within the network, subscribers were frequency-division multiplexed across shared physical facilities, thus allowing the telephone company to efficiently conserve network bandwidth.

Time, of course, changes everything. As with any technology, frequency-division multiplexing has its downsides. It is an analog technology and therefore suffers from the shortcomings that have historically plagued all transmission systems. The wire over which information is transmitted behaves like a long-wire antenna, picking up noise along the length of the transmission path and effectively homogenizing it with the voice signal. Additionally, the power of the transmitted signal diminishes over distance, and if the distance is far enough, the signal will have to be amplified to make it intelligible at the receiving end. Unfortunately, the amplifiers used in the network are not particularly discriminating: they have no way of separating the voice noise. The result is that they convert a weak, noisy signal into a loud, noisy signal. This is better, but far from ideal. A better solution was needed.

The better solution came about with the development of Time-Division Multiplexing (TDM), which became possible because of the transistor and integrated circuit electronics that arrived in the late 1950s and early 1960s. TDM is a digital transmission scheme, which implies a small number of discrete signal states, rather than the essentially infinite range of values employed in analog systems (the word *digital* literally means *discrete*). Although digital systems are just as susceptible to noise impairment as their analog counterparts, the discrete nature of their binary signaling makes it relatively easy to separate the noise from the transmitted signal. Digital carrier systems have only three valid signal values: one positive, one negative, and zero; anything else is construed to be noise. It is therefore a trivial exercise for digital repeaters to discern what is desirable and what is not, thus eliminating the problem of cumulative noise. The role of the regenerator, shown in Figure 1-16, is to receive a weak, noisy digital signal, remove the noise, reconstruct the original signal, and amplify it before transmitting the signal onto the next segment of the transmission facility. For this reason, repeaters are also called *regenerators* because that is precisely the function they perform.

One observation: it is estimated that as much as 60 percent of the cost of building a transmission facility lies in the regenerator sections of the span. For this reason, optical networking, discussed a bit later, has various benefits, not the least of which is the capability to reduce the number of regenerators required on long transmission spans. In a typical network, these regenerators must be placed approximately every 6,000 feet along a span, which means that considerable expense is involved when providing regeneration along a long-haul network.

Digital signals, often called square waves, comprise a very rich mixture of signal frequencies. Not to bring too much physics into the discussion, but we must at least mention the Fourier series, which describes the makeup of a digital signal. The Fourier series is a mathematical representation of the behavior of waveforms. Among other things, it notes the following fact. If we start with a fundamental signal such as that shown in Figure 1-17, and

Figure 1-16
Regeneration of a
weak, noisy signal.

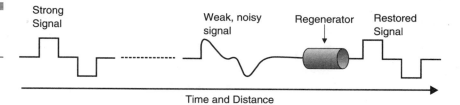

Figure 1-17
Analog to digital
conversion.

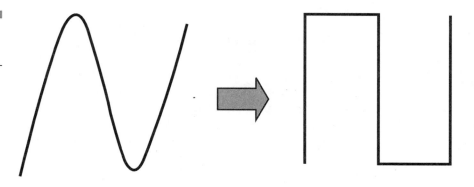

mathematically add to it its odd harmonics (a harmonic is defined as a wave whose frequency is a whole-number multiple of another wave), we see a rather remarkable thing happening: the waveform becomes steeper on the sides and flatter on top. As we add more and more of the odd harmonics (there is, after all, an infinite series of them), the wave begins to look like the typical square wave. Now of course, there is no such thing as a true square wave; for our purposes, though, we'll accept the fact.

It should now be intuitive to the reader that digital signals comprise a mixture of low, medium, and high frequency components, which means that they cannot be transmitted across the bandwidth-limited 4 KHz channels of the traditional telephone network. In digital carrier facilities, the equipment that restricts the individual transmission channels to 4-KHz chunks is eliminated, thus giving each user access to the full breadth of available spectrum across the shared physical medium. In frequency division systems, we observed that we give users "some of the frequency all of the time;" in time-division systems, we turn that around and give users "*all* of the frequency *some* of the time." As a result, high-frequency digital signals can be transmitted without restriction.

Digitization brings a cadre of advantages, including improved voice and data transmission quality; better maintenance and troubleshooting capability, and therefore reliability; and dramatic improvements in configuration flexibility. In digital carrier systems, the time-division multiplexer is known as a channel bank; under normal circumstances, it enables either 24 or 30 circuits to share a single, four-wire facility. The 24-channel system is called T-Carrier; the 30-channel system, used in most of the world, is called E-Carrier. Originally designed in 1962 as a way to transport multiple channels of voice over expensive transmission facilities, they soon became useful as data transmission networks as well. That, however, came later. For now, we focus on voice.

Voice Digitization

The process of converting analog voice to a digital representation in the modern network is a logical and straightforward process. It comprises four distinct steps: *Pulse Amplitude Modulation* (PAM) sampling, in which the amplitude of the incoming analog wave is sampled every 125 microseconds; *companding*, during which the values are weighted toward those most receptive to the human ear; *quantization*, in which the weighted samples are given values on a nonlinear scale; and finally *encoding*, during which each value is assigned a distinct binary value. Each of these stages of *Pulse Code Modulation* (PCM) will now be discussed in detail.

Pulse Code Modulation (PCM)

Thanks to work performed by Harry Nyquist at Bell Laboratories in the 1920s, we know that to optimally represent an analog signal as a digitally encoded bitstream, the analog signal must be sampled at a rate that is equal to twice the bandwidth of the channel over which the signal is to be transmitted. Because each analog voice channel is allocated 4 KHz of bandwidth, it follows that each voice signal must be sampled at twice that rate, or 8,000 samples per second. In fact, that is precisely what happens in T-Carrier systems, which we will use to illustrate our example. The standard T-Carrier multiplexer accepts inputs from 24 analog channels, as shown in Figure 1-18. Each channel is sampled in turn, every one eight-thousandth of a second in round-robin fashion, resulting in the generation of 8,000 pulse amplitude samples from each channel every second. The sampling rate is important. If the sampling rate is too high, too much information is transmitted and bandwidth is wasted; if the sampling rate is too low, then we run the risk of aliasing. Aliasing is the interpretation of the sample points as a false waveform, due to the paucity of samples.

This Pulse Amplitude Modulation process represents the first stage of Pulse Code Modulation, the process by which an analog baseband signal is converted to a digital signal for transmission across the T-Carrier network. Figure 1-19 shows this first step.

The second stage of PCM, shown in Figure 1-20, is called quantization. In quantization, we assign values to each sample within a constrained range.

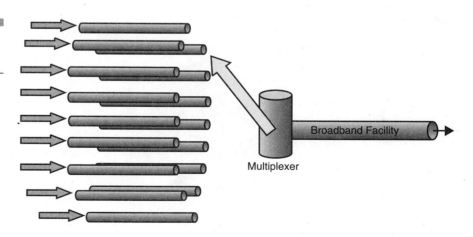

Figure 1-18
Time division
multiplexing.

Broadband Facility

Multiplexer

Inbound Traffic

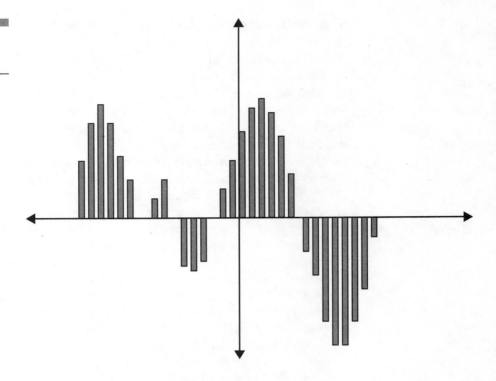

Figure 1-19
*Pulse Amplitude
Modulation (PAM).*

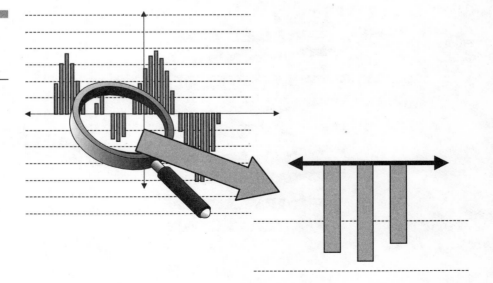

Figure 1-20
*Quantization
and Pulse Code
Modulation (PCM).*

For illustration purposes, imagine what we now have before us. We have replaced the continuous analog waveform of the signal with a series of amplitude samples that are close enough together that we can discern the shape of the original wave from their collective amplitudes. Imagine also that we have graphed these samples in such a way that the wave of sample points meanders above and below an established zero point on the x-axis, so that some of the samples have positive values and others are negative.

The amplitude levels enable us to assign values to each of the PAM samples, although a glaring problem with this technique should be obvious to the careful reader. Very few of the samples actually line up *exactly* with the amplitudes delineated by the graphing process. In fact, most of them fall between the values, as shown in the illustration. It doesn't take much of an intuitive leap to see that several of the samples will be assigned the same digital value by the coder-decoder that performs this function, yet they are clearly not the same amplitude. This inaccuracy in the measurement method results in a problem known as *quantizing noise* and is inevitable when linear measurement systems, such as the one suggested by the drawing, are employed in *coder-decoders* (CODECs).

Needless to say, design engineers recognized this problem rather quickly, and came up with an adequate solution just as quickly. It is a fairly well-known fact among psycholinguists and speech therapists that the human ear is far more sensitive to discrete changes in amplitude at low-volume levels than it is at high-volume levels, a fact not missed by the network designers tasked with optimizing the performance of digital carrier systems intended for voice transport. Instead of using a linear scale for digitally encoding the PAM samples, they designed and employed a nonlinear scale that is weighted with much more granularity at low-volume levels (that is, close to the zero line) than at the higher amplitude levels. In other words, the values are extremely close together near the x-axis, and become farther and farther apart as they travel up and down the y-axis. This nonlinear approach keeps the quantizing noise to a minimum at the low amplitude levels where hearing sensitivity is the highest, and enables it to creep up at the higher amplitudes, where the human ear is less sensitive to its presence. It turns out that this is not a problem because the inherent shortcomings of the mechanical equipment (microphones, speakers, the circuit itself) introduce slight distortions at high amplitude levels that hide the effect of the nonlinear quantizing scale.

This technique of compressing the values of the PAM samples to make them fit the nonlinear quantizing scale results in bandwidth savings of more than 30 percent. The actual process is called companding because the sample is first compressed for transmission, then expanded for reception at the far end.

The actual graph scale is divided into 255 distinct values above and below the zero line. In North America and Japan, the encoding scheme is known as μ-Law (Mu-Law); the rest of the world relies on a slightly different standard known as A-Law.

Eight segments are above the line and eight are below (one of which is the shared zero point); each segment, in turn, is subdivided into 16 steps. A bit of binary mathematics now enables us to convert the quantized amplitude samples into an eight-bit value for transmission. For the sake of demonstration, let's consider a negative sample that falls into the thirteenth step in segment five. The conversion would take on the following representation

1 101 1101

where the initial 0 indicates a negative sample, 101 indicates the fifth segment, and 1101 indicates the thirteenth step in the segment. We now have an eight-bit representation of an analog amplitude sample that can be transmitted across a digital network, then reconstructed with its many counterparts as an accurate representation of the original analog waveform at the receiving end. This entire process is known as Pulse Code Modulation (PCM) and the result of its efforts is often referred to as toll-quality voice.

Alternative Digitization Techniques

Although PCM is perhaps the best-known, high-quality voice digitization process, it is by no means the only one. Advances in coding schemes and improvements in the overall quality of the telephone network have made it possible to develop encoding schemes that use far less bandwidth than traditional PCM. In this next section, we will consider some of these techniques.

Adaptive Differential Pulse Code Modulation (ADPCM)   *Adaptive Differential Pulse Code Modulation* (ADPCM) is a technique that enables toll-quality voice signals to be encoded at half-rate (32 Kbps) for transmission. ADPCM relies on the predictability that is inherent in human speech to reduce the amount of information required. The technique still relies on PCM encoding, but adds an additional step to carry out its task. The 64 Kbps PCM-encoded signal is fed into an ADPCM transcoder, which considers the previous behavior of the incoming stream to create a prediction of the behavior of the next sample. This is where the magic happens: instead of transmitting the actual value of the predicted sample, it encodes in four bits and transmits the *difference* between the actual and predicted samples. Because the difference from sample to sample is typically quite small, the results are generally considered to be very close to toll-quality. This four-bit transcoding process, which is based on the known behavior characteristics of human voice, enables the system to transmit 8,000 four-bit samples per second, thus reducing the overall bandwidth requirement from 64 Kbps to 32 Kbps. It should be noted that ADPCM works well for voice because the encoding and predictive algorithms are based upon its behavior characteristics. It does not, however, work as well for higher bit rate data (above 4,800 bps), which has an entirely different set of behavior characteristics.

Continuously Variable Slope Delta (CVSD) *Continuously Variable Slope Delta* (CVSD) is a unique form of voice encoding that relies on the values of individual bits to predict the behavior of the incoming signal. Instead of transmitting the volume (height or y-value) of PAM samples, CVSD transmits information that it measures the changing slope of the waveform. Rather than transmitting the actual change itself, it transmits the *rate* of change, as shown in Figure 1-21.

To perform its task, CVSD uses a reference voltage to which it compares all incoming values. If the incoming signal value is less than the reference voltage, then the CVSD encoder reduces the slope of the curve to make its approximation better mirror the slope of the actual signal. If the incoming

Figure 1-21
Continuously
Variable Slope
Delta Modulation.

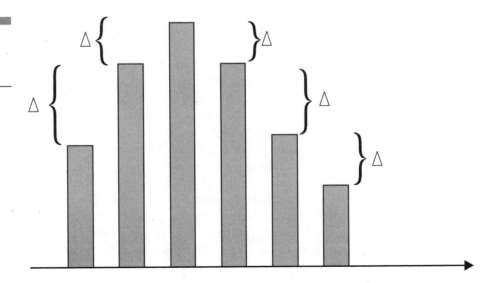

value is more than the reference value, then the encoder will increase the slope of the output signal, again causing it to approach and therefore mirror the slope of the actual signal. With each recurring sample and comparison, the step function can be increased or decreased as required. For example, if the signal is increasing rapidly, then the steps are increased one after the other in a form of step function by the encoding algorithm. Obviously, the reproduced signal is not a particularly exact representation of the input signal: in practice, it is pretty jagged. Filters, therefore, are used to smooth the transitions.

CVSD is typically implemented at 32 Kbps, although it can be implemented at rates as low as 9,600 bps. At 16–24 Kbps, recognizability is still possible; down to 9,600, recognizability is seriously affected, although intelligibility is not.

Linear Predictive Coding (LPC)   We mention *Linear Predictive Coding* (LPC) here only because it has carved out a niche for itself in certain voice-related applications such as voice mail systems, automobiles, aviation, and electronic games that speak to children. LPC is a complex process, implemented completely in silicon, which enables voice to be encoded at rates as low as 2,400 bps. The resulting quality is far from toll-quality, but it is certainly intelligible and its low-bit rate capability gives it a distinct advantage over other systems.

Linear Predictive Coding relies on the fact that each sound created by the human voice has unique attributes, such as frequency range, resonance, and loudness, among others. When voice samples are created in LPC, these attributes are used to generate prediction coefficients. These predictive coefficients represent linear combinations of previous samples, hence the name, Linear Predictive Coding.

Prediction coefficients are created by taking advantage of the known *formants* of speech, which are the resonant characteristics of the mouth and throat that give speech its characteristic timbre and sound. This sound, referred to by speech pathologists as the *buzz*, can be described by both its pitch and its intensity. LPC, therefore, models the behavior of the vocal cords and the vocal tract itself.

To create the digitized voice samples, the buzz is passed through an inverse filter that is selected based upon the value of the coefficients. The remaining signal, after the buzz has been removed, is called the residue.

In the most common form of LPC, the residue is encoded as either a *voiced* or *unvoiced* sound. Voiced sounds are those that require vocal cord vibration, such as the *g* in *glare,* the *b* in *boy,* the *d* and *g* in *dog.* Unvoiced sounds require no vocal cord vibration, such as the *h* in *how,* the *sh* in *shoe,* and the *f* in *frog*. The transmitter creates and sends the prediction coefficients, which include measures of pitch, intensity, and whatever voiced and unvoiced coefficients that are required. The receiver undoes the process; it converts the voice residue, pitch, and intensity coefficients into a representation of the source signal, using a filter similar to the one used by the transmitter to synthesize the original signal.

Digital Speech Interpolation (DSI) Human speech has many measurable (and therefore predictable) characteristics, one of which is a tendency to have embedded pauses. As a rule, people do not spew out a series of uninterrupted sounds; they tend to pause for emphasis, to collect their thoughts, and to reword a phrase while the other person listens quietly on the other end of the line. When speech technicians monitor these pauses, they discover that during considerably more than half of the total connect time, the line is silent.

Digital Speech Interpolation (DSI) takes advantage of this characteristic silence to drastically reduce the bandwidth required for a single channel. Whereas 24 channels can be transported over a typical T-1 facility, DSI enables as many as 120 conversations to be carried over the same circuit. The format is proprietary and requires the setting aside of a certain amount of bandwidth for overhead.

A form of statistical multiplexing lies at the heart of DSI's functionality. Standard T-Carrier is a time-division multiplexed scheme, in which channel ownership is assured: a user assigned to channel three will *always* own channel three, regardless of whether he or she is actually using the line. In DSI, channels are not owned. Instead, large numbers of users share a pool of available channels. When a user starts to talk, the DSI system assigns an available timeslot to that user and notifies the receiving end of the assignment. This system works well when the number of users is large because statistical probabilities are more accurate and indicative of behavior in larger populations than in smaller ones.

DSI has a downside, of course, and it comes in several forms. *Competitive clipping* occurs when more people start to talk than there are available channels, resulting in someone being unable to talk. *Connection clipping* occurs when the receiving end fails to learn what channel a conversation has been assigned within a reasonable amount of time, resulting in signal loss. Two approaches have been created to address these problems; in the case of competitive clipping, the system intentionally clips off the front end of the initial word of the second person who speaks. This technique is not optimal, but does prevent loss of the conversation and also obviates the problem of clipping out the middle of a conversation, which would be more difficult for the speakers to recover from. The loss of an initial syllable or two can be mentally reconstructed far more easily than sounds in the middle of a sentence.

A second technique used to recover from clipping problems is to temporarily reduce the encoding rate. The typical encoding rate for DSI is 32 Kbps; in certain situations, the encoding rate may be reduced to 24 Kbps, thus freeing up significant bandwidth for additional channels. Both techniques are widely utilized in DSI systems.

Framing and Formatting in T-1

The standard T-Carrier multiplexer accepts inputs from 24 sources, converts the inputs to PCM bytes, then time-division multiplexes the samples over a shared four-wire facility, as shown in Figure 1-22. Each of the 24 input channels yields an eight-bit sample, in round-robin fashion, once every 125 microseconds (8,000 times per second). This yields an overall bit rate of 64 Kbps for each channel (eight bits per sample/8,000 samples per second). The multiplexer gathers one eight-bit sample from each of the 24 channels, and aggregates them into a 192-bit frame. To the frame, it adds a frame bit, which expands the frame to a 193-bit entity. The frame bit is used for a variety of purposes that will be discussed in a moment.

Figure 1-22
Creating a T-Carrier frame.

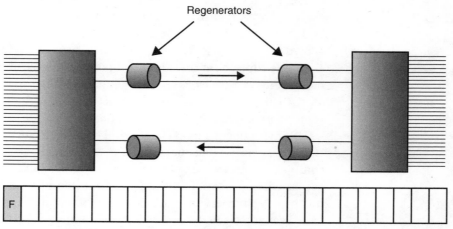

8-bits per sample, 24 samples per frame + frame bit = 193 bits.
8,000 frames are generated per second, yielding 1.544 Mbps.

The 193-bit frames of data are transmitted across the four-wire facility at the standard rate of 8,000 frames per second, for an overall T-1 bit rate of 1.544 Mbps. Keep in mind that 8 Kbps of the bandwidth consist of frame bits (one frame bit per frame, 8,000 frames per second); only 1.536 Mbps belong to the user.

Beginnings: D1 Framing

The earliest T-Carrier equipment was referred to as D1 and was considerably more rudimentary in function than modern systems (see Figure 1-23). In D1, every eight-bit sample carried seven bits of user information (bits one through seven) and one bit for signaling (bit eight). The signaling bits were used for exactly that: indications of the status of the line (on-hook, off-hook, busy, high and dry, and so on), whereas the seven user bits carried encoded voice information. Because only seven of the eight bits were available to the user, the result was considered to be less than toll quality (128 possible values, rather than 256). The frame bits, which in modern systems indicate the beginning of the next 192-bit frame of data, toggled back and forth between zero and one.

Figure 1-23
D1 signaling.

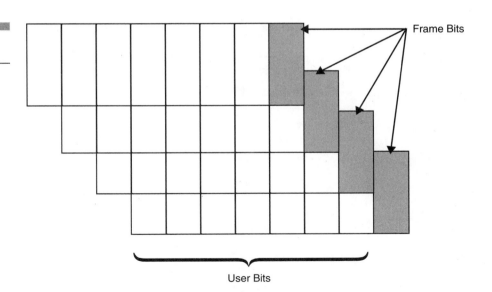

Frame Bits

User Bits

Evolution: D4

As time went on and the stability of network components improved, an improvement on D1 was sought after and found. Several options were developed, but the winner emerged in the form of the D4 or superframe format. Rather than treat a single 193-bit frame as the transmission entity, superframe gangs together 12 193-bit frames into a 2,316-bit entity, shown in Figure 1-24, that obviously includes 12 frame bits. Please note that the bit rate has not changed; we have simply changed our view of what constitutes a frame.

Because we now have a single (albeit large) frame, we clearly don't need 12 frame bits to frame it; consequently, some of them can be redeployed for other functions. In superframe, the six odd-numbered frame bits are referred to as terminal-framing bits and are used to synchronize the channel bank equipment. The even-framing bits, on the other hand, are called signal-framing bits and are used to indicate to the receiving device where robbed-bit signaling occurs.

In D1, the system reserved one bit from every sample for its own signaling purposes, which succeeded in reducing the user's overall throughput. In D4, that is no longer necessary; instead, we signal less frequently, and only occasionally rob a bit from the user. In fact, because the system operates at a high transmission speed, network designers determined that signaling

Figure 1-24
Superframe (SF).

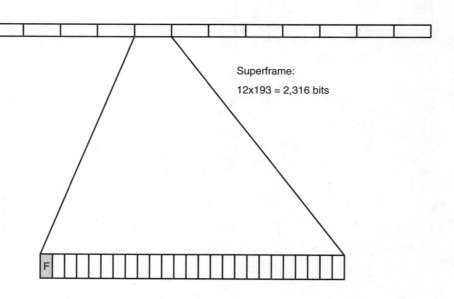

Superframe:

12x193 = 2,316 bits

can occur relatively infrequently and still convey adequate information to the network. Consequently, bits are robbed from the sixth and eighth iteration of each channel's samples and then only the least significant bit from each sample. The resulting change in voice quality is negligible.

Back to the signal-framing bits: within a transmitted superframe, the second and fourth signal-framing bits would be the same, but the sixth would toggle to the opposite value, indicating to the receiving equipment that the samples in that subframe of the superframe should be checked for signaling state changes. The eighth and tenth signal-framing bits would stay the same as the sixth, but would toggle back to the opposite value once again in the twelfth, indicating once again that the samples in that subframe should be checked for signaling state changes.

Today: Extended Superframe (ESF)

Although superframe continues to be widely utilized, an improvement came about in the 1980s in the form of *extended superframe* (ESF), shown in Figure 1-25. ESF groups 24 frames into an entity instead of 12, and like superframe, it reuses some of the frame bits for other purposes. Bits 4, 8, 12, 16, 20, and 24 are used for framing, and form a constantly repeating pattern (001011 . . .). Bits 2, 6, 10, 14, 18, and 22 are used as a six-bit *cyclic redundancy check* (CRC) to check for bit errors on the facility.

Figure 1-25
Extended Superframe
(ESF).

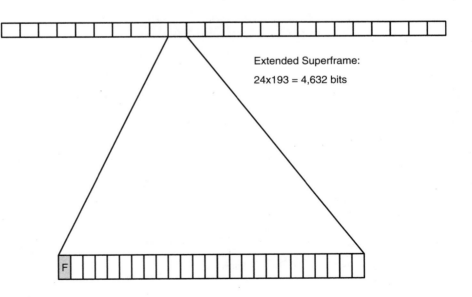

Extended Superframe:

24x193 = 4,632 bits

Finally, the remaining bits, all of the odd frame bits in the frame, are used as a 4 Kbps facility data link for end-to-end diagnostics and network management tasks.

ESF provides one major benefit over its predecessors: the capability to do non-intrusive testing of the facility. In earlier systems, if the user reported trouble on the span, the span would have to be taken out of service for testing. With ESF, that is no longer necessary because of the added functionality provided by the CRC and the facility data link.

The Rest of the World: E-1

E-1, used for the most part outside of the U.S. and Canada, differs from T-1 on several key points. First, it boasts a 2.048 Mbps facility, rather than the 1.544 Mbps facility found in T-1. Second, it utilizes a 32-channel frame rather than 24. Channel one contains framing information and a *four-bit cyclic redundancy check* (CRC-4); Channel 16 contains all signaling information for the frame; and channels one through 15 and 17 through 31 transport user traffic. Figure 1-26 shows the frame structure.

A number of similarities exist between T-1 and E-1 as well: channels are all 64 Kbps and frames are transmitted 8,000 times per second. Whereas T-1 gangs together 24 frames to create an extended superframe, E-1 gangs together 16 frames to create what is known as an ETSI multiframe. The multiframe is subdivided into two sub-multiframes; the CRC-4 in each one is used to check the integrity of the sub-multiframe that preceded it.

A final word about T-1 and E-1: because T-1 is a departure from the international E-1 standard, it is incumbent upon the T-1 provider to perform all interconnection conversions between T-1 and E-1 systems. For example, if a

Figure 1-26
An E-1 frame showing frame, signaling bits.

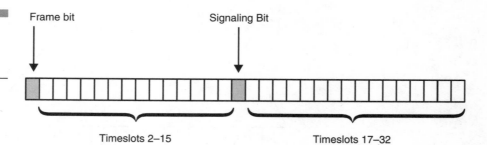

Frame bit

Signaling Bit

Timeslots 2–15

Timeslots 17–32

call arrives in the U.S. from a European country, the receiving American carrier must convert the incoming E-1 signal to T-1. If a call originates from Canada and is terminated in Australia, the Canadian originating carrier must convert the call to E-1 before transmitting it to Australia.

Up the Food Chain: From T-1 to DS3 . . . and Beyond

When T-1 and E-1 first emerged on the telecommunications scene, they represented a dramatic step forward in terms of the bandwidth that service providers now had access to. In fact, they were so bandwidth rich that there was no concept that a customer would ever need access to them. What customer, after all, could ever have a use for 1.5 million bits per second of bandwidth?

Of course, that question was rendered moot in short order as increasing requirements for bandwidth drove demand that went well beyond the limited capabilities of low-speed transmission systems. As T-1 became more mainstream, its usage went up, and soon requirements emerged for digital transmission systems with capacity greater than 1.544 Mbps. The result was the creation of what came to be known as the North American Digital Hierarchy, shown in Figure 1-27. The table also shows the European and Japanese hierarchy levels.

Figure 1-27
International Multiplexing hierarchies.

Hierarchy Level	Europe	United States	Japan
DS-0	64 Kbps	64 Kbps	64 Kbps
DS-1		1.544 Mbps	1.544 Mbps
E-1	2.048 Mbps		
DS-1c		3.152 Mbps	3.152 Mbps
DS-2		6.312 Mbps	6.312 Mbps
E-2	8.448 Mbps		32.064 Mbps
DS-3	34.368 Mbps	44.736 Mbps	
DS-3c		91.053 Mbps	
E-3	139.264 Mbps		
DS-4		274.176 Mbps	
			397.2 Mbps

From DS-1 to DS-3

We have already seen the process employed to create the DS-1 signal from 24 incoming DS-0 channels and an added frame bit. Now we turn our attention to higher bit rate services. As we wander our way through this explanation, pay particular attention to the complexity involved in creating higher rate payloads. This is one of the great advantages of SONET and SDH.

The next level in the North American Digital Hierarchy is called DS-2. Although it is rarely seen outside of the safety of the multiplexer in which it resides, it plays an important role in the creation of higher bit rate services. It is created when a multiplexer *bit interleaves* four DS-1 signals, inserting as it does so a control bit, known as a C-bit, every 48 bits in the payload stream. Bit interleaving is an important construct because it contributes to the complexity of the overall payload. In a bit interleaved system, multiple bit streams are combined on a bit-by-bit basis, as shown in Figure 1-28. When payload components are bit-interleaved to create a higher rate multiplexed signal, the system first selects bit one from channel one, bit one from channel two, bit one from channel three, and so on. Once it has selected and transmitted all of the first bits, it goes on to the second bits from each channel, then the third, until it has created the super-rate frame. Along the way it intersperses C-bits, which are used to perform certain control and management functions within the frame.

Once the 6.312 Mbps DS-2 signal has been created, the system shifts into high gear to create the next level in the transmission hierarchy. Seven DS-2 signals are then bit-interleaved along with C-bits after every 84 payload bits to create a composite 44.736 Mbps DS-3 signal. The first part of this

Figure 1-28
Bit interleaving.

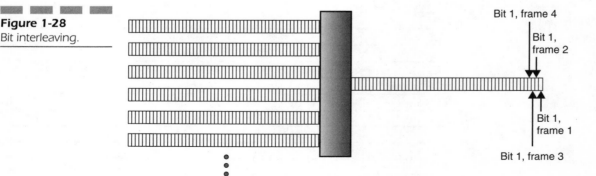

Bit 1, frame 4

Bit 1, frame 2

Bit 1, frame 1

Bit 1, frame 3

process, the creation of the DS-2 payload, is called *M12 multiplexing*; the second step, which combines DS-2s to form a DS-3, is called *M23 multiplexing*. The overall process is called *M13*, and is illustrated in Figure 1-29.

The problem with this process is the bit-interleaved nature of the multiplexing scheme. Because the DS-1 signal components arrive from different sources, they may be (and usually are) slightly off from one another in terms of the overall phase of the signal; in effect, their speeds differ slightly. This is unacceptable to a multiplexer, which must rate-align them if it is to properly multiplex them, beginning with the head of each signal. In order to do this, the multiplexer inserts additional bits, known as stuff bits, into the signal pattern at strategic places that serve to rate align the components. The structure of a bit-stuffed DS-2 frame is shown in Figure 1-30; a DS-3 frame is shown in Figure 1-31.

Figure 1-29
The M13 multiplexing process.

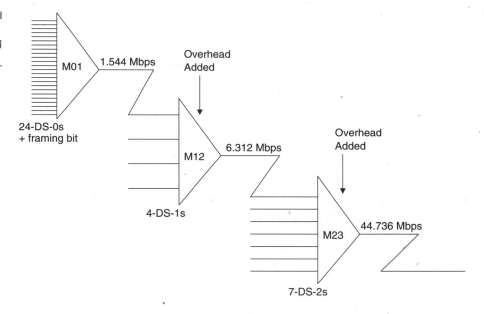

Figure 1-30
M12 frame comprises four sub-frames and 48-bit payload fields: 1,176 bits.

M0		C1		F0		C1		C1		F1	
M1		C2		F0		C2		C2		F1	
M1		C3		F0		C3		C3		F1	
M1		C4		F0		C4		C4		F1	

Figure 1-31
M13 frame comprises
seven sub-frames and
84-bit payload fields:
4,760 bits.

X1		F1		C1		F0		C2		F0		C3		F1	
X1		F1		C1		F0		C2		F0		C3		F1	
P1		F1		C1		F0		C2		F0		C3		F1	
P2		F1		C1		F0		C2		F0		C3		F1	
M1		F1		C1		F0		C2		F0		C3		F1	
M2		F1		C1		F0		C2		F0		C3		F1	
M3		F1		C1		F0		C2		F0		C3		F1	

The complexity of this process should now be fairly obvious to the reader. If we follow the left-to-right path shown in Figure 1-32, we see the rich complexity that suffuses the M13 signal-building process. Twenty-four 64 Kbps DS0s are aggregated at the ingress side of the T-1[1] multiplexer, grouped into a T-1 frame, and combined with a single frame bit to form an outbound 1.544 Mbps signal (we call this the M01 stage; that's our nomenclature, used for the sake of naming continuity). That signal then enters the intermediate M12 stage of the multiplexer, where it is combined (bit-interleaved) with three others and a good dollop of alignment overhead to form a 6.312 Mbps DS-2 signal. That DS-2 then enters the M23 stage of the mux, where it is bit-interleaved with six others and another scoop of overhead to create a DS-3 signal. At this point, we have a relatively high-bandwidth circuit that is ready to be moved across the wide area network.

Of course, as our friends in the U.K. are wont to say, the inevitable spanner is always tossed into the works (those of us on the left side of the Atlantic call it a wrench). Keep in mind that the 28 (do the math) bit-interleaved DS-1s may well come from 28 different sources, which means that they may well have 28 different destinations. This translates into the pre-SONET digital hierarchy's greatest weakness and one of SONET's greatest advantages. In order to drop a DS-1 at its intermediate destination, we have to bring the composite DS-3 into a set of back-to-back DS-3 multiplexers (sometimes called M13 multiplexers). There, the ingress mux removes the second set of overhead, finds the DS-2 in which the DS-1 we have to drop out is carried, removes its overhead, finds the right DS-1, drops it out, then rebuilds the DS-3 frame, including reconstruction of the overhead, before transmitting it on to its next destination. This process is complex, time-consuming, and expensive. So what if we could come up with a

---

[1] The process is similar for the E-1 hierarchy.

method for adding and dropping signal components that eliminated the M13 process entirely? What if we could do it as simply as the process shown in Figure 1-33?

Figure 1-32
The complexity of M13.

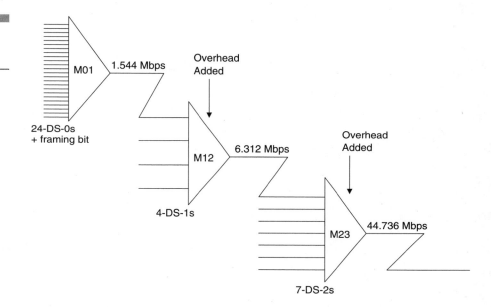

24-DS-0s + framing bit

M01   1.544 Mbps

Overhead Added

M12   6.312 Mbps

4-DS-1s

Overhead Added

M23   44.736 Mbps

7-DS-2s

Figure 1-33
Dropping a payload component in the M13 environment.

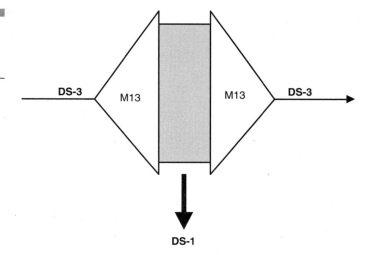

DS-3   M13   M13   DS-3

DS-1

We have. It's called SONET in North America, SDH in the rest of the world, and it dramatically simplifies the world of high-speed transport. How does it do this? That's the subject of Chapter Two.

SONET Basics

Before we descend into the technological depths of the SONET frame structure, let's revisit the purpose for SONET in the first place. Remember that the digital hierarchy (DS-0, DS-1, DS-2, DS-3, and so on) was created to provide cost-effective multiplexed transport for voice and data traffic from one location in a network to another. SONET has the same responsibility, albeit on a larger scale: indeed, it is sometimes described as "T-1 on steroids."

SONET brings with it a subset of advantages that makes it stand above competitive technologies. These include mid-span meet, improved *operations, administration, maintenance, and provisioning* (OAM&P), support for multipoint circuit configurations, non-intrusive facility monitoring, and the ability to deploy a variety of new services. We will examine each of these in the following sections.

Mid-Span Meet

When Bill McGowan first complained about the fact that Equal Access required him to ultimately spend MCI's money on AT&T hardware, he was complaining about the lack of a *mid-span meet* ability—in other words, the ability of one vendor's optical multiplexer to not only connect electrically with that of another vendor, but to actually pass understandable maintenance messages between the two. Because of the monopoly nature of early networks, interoperability was a laughable dream. Following the divestiture of AT&T, however, and the realization of Equal Access, the need for interoperability standards became a matter of some priority. Mid-span meet was SONET's contribution to this important effort.

Improved OAM&P

Improved OAM&P is without question one of the greatest contributions that SONET brings to the networking table. Element and network monitoring, management, and maintenance have always been something of a catch-as-catch-can effort because of the complexity and diversity of elements in a typical service provider's network. SONET overhead includes error-checking ability, bytes for network survivability, and a diverse set of clearly defined management messages.

Multipoint Circuit Support

When SONET was first deployed in the network, the bulk of the traffic it carried derived from point-to-point circuits such as T-1 and DS-3 facilities. With SONET came the ability to hub the traffic, a process that combines the best of cross-connection and multiplexing to perform a ability known as *groom and fill*. This means that aggregated traffic from multiple sources can be transported to a hub, managed as individual components, and redirected out any of several outbound paths without having to completely disassemble the aggregate payload. Prior to SONET, this process required a pair of back-to-back multiplexers, sometimes called an M13 (for a multiplexer that interfaces between DS-1 and DS-3). This ability, combined with SONET's discreet and highly capable management features, results in a wonderfully manageable system of network bandwidth control.

Non-Intrusive Monitoring

SONET overhead bytes are embedded in the frame structure, meaning that they are universally transported alongside the customer's payload. Thus, tight and granular control over the entire network can be realized, leading to more efficient network management and the ability to deploy services on an as-needed basis.

New Services

SONET bandwidth is imminently scalable, meaning that the ability to provision additional bandwidth for customers that require it on an as-needed basis becomes real. As applications evolve to incorporate more and more multimedia content and to therefore require greater volumes of bandwidth, SONET offers it by the bucket load. Already, interfaces between SONET and Gigabit Ethernet are being written; interfaces to ATM and other high-speed switching architectures have been in existence for some time already.

SONET Evolution

SONET was initially designed to provide multiplexed point-to-point transport. However, as its capabilities became better understood and networks became mission-critical, its deployment became more innovative, and soon it was deployed in ring architectures, as shown in Figure 2-1. These rings, which are described later, represent one of the most commonly deployed network topologies. For the moment, however, let's examine a point-to-point deployment. As it turns out, rings don't differ all that much.

If we consider the structure and function of the typical point-to-point circuit, we find a variety of devices and *functional regions*, as shown in Figure 2-2. The components include end-devices, multiplexers in this case, which provide the point of entry for traffic originating in the customer's equipment and seeking transport across the network; a full-duplex circuit, which provides simultaneous two-way transmission between the network components; a series of repeaters or regenerators, responsible for periodically reframing and regenerating the digital signal; and one or more intermediate multiplexers, which serve as nothing more than pass-through devices.

When non-SONET traffic is transmitted into a SONET network, it is packaged for transport through a step-by-step, quasi-hierarchical process

Figure 2-1
SONET ring
architectures.

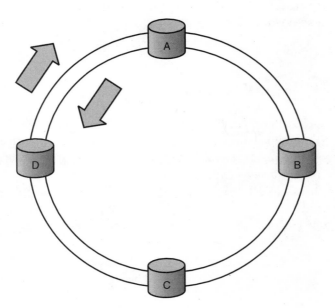

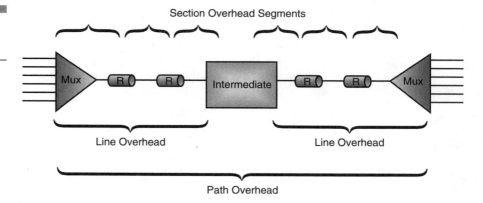

Figure 2-2
SONET overhead in
the network.

that attempts to make reasonably good use of the available network bandwidth and ensure that receiving devices can interpret the data when it arrives. The intermediate devices, including multiplexers and repeaters, also play a role in guaranteeing traffic integrity, and to that end, the SONET standards divide the network into three regions: path, line, and section. To understand the differences between the three, let's follow a typical transmission of a DS-3, probably carrying 28 T-1s, from its origination point to the destination.

When the DS-3 first enters the network, the ingress SONET multiplexer packages it by wrapping it in a collection of additional information, called *Path Overhead*, which is unique to the transported data. For example, it attaches information that identifies the original source of the DS-3, so that it can be traced in the event of network transmission problems; a bit-error control byte; information about how the DS-3 is actually mapped into the payload-transport area (and unique to the payload type); an area for network performance and management information; and a number of other informational components that have to do with the end-to-end transmission of the unit of data.

The packaged information, now known as a *payload,* is inserted into a SONET frame, and at that point, another layer of control and management information is added, called *Line Overhead.* Line Overhead is responsible for managing the movement of the payload from multiplexer to multiplexer. To do this, it adds a set of bytes that enable receiving devices to find the payload inside the SONET frame. As you will learn a bit later, the payload can occasionally wander around inside the frame due to the vagaries of the network. These bytes enable the system to track that movement.

In addition to these tracking bytes, the Line Overhead includes bytes that monitor the integrity of the network and have the ability to effect a switch to a backup transmission span if a failure in the primary span occurs. It also includes another bit-error checking byte, a robust channel for transporting network-management information, and a voice communications channel that enables technicians at either end of the line to plug in with a handset (sometimes called a butt-in or buttinski) and communicate while troubleshooting.

The final step in the process is to add a layer of overhead that enables the intermediate repeaters to find the beginning of and synchronize a received frame. This overhead, called the *Section Overhead,* contains a unique initial-framing pattern at the beginning of the frame, an identifier for the payload signal being carried, another bit-error check, a voice communications channel, and another dedicated channel for network management information, similar to but smaller than the one identified in the Line Overhead.

The result of all this overhead, much of which seems like overkill (and in many peoples' minds it is), is that the transmission of a SONET frame containing user data can be identified and managed with tremendous granularity from the source all the way to the destination.

So, to summarize, the hard little kernel of DS-3 traffic is gradually surrounded by three layers of overhead information, as shown in Figure 2-3, that help it achieve its goal of successfully transiting the network. The Section Overhead is used at every device the signal passes through, including multiplexers and repeaters; the Line Overhead is only used between multiplexers; and the information contained in the Path Overhead is only used by the source and destination multiplexers—the intermediate multiplexers don't care about the specific nature of the payload because they don't have to terminate or interpret it.

One final point for the protocol purists out there: the SONET overhead is often described as shown in Figure 2-4. This layered model can be a bit misleading because it seems to imply a hierarchy of functionality or intelligence. Make no mistake about it, though: SONET is purely a physical-layer standard. A functional hierarchy of sorts may exist among the three overhead types, but they are all mired in the primordial ooze of the transmission layer.

Enough about protocol. On to the SONET frame structure.

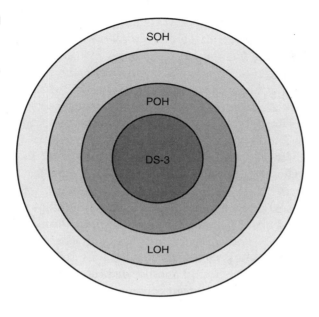

Figure 2-3
The layers of the
SONET overhead.

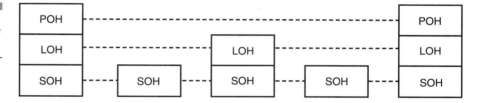

Figure 2-4
The layered nature of
SONET overhead.

The SONET Frame

Keep in mind once again that we are doing nothing more complicated than building a T-1 frame with an attitude. Recall that the T-1 frame comprised 24 eight-bit channels (samples from each of 24 incoming data streams) plus a single bit of overhead. In SONET, we have a similar construct—much more channel capacity and much more overhead is available, but it has the same functional concept.

Figure 2-5 shows the fundamental SONET frame. This frame is known as a *Synchronous Transport Signal, Level One* (STS-1). It is 9 bytes tall and 90 bytes wide for a total of 810 bytes of transported data including both user payload and overhead. The first three columns of the frame are the Section and Line Overhead, known collectively as the *Transport Overhead*. The bulk of the frame itself, to the left, is the *synchronous payload envelope* (SPE), which is the container area for the user data that is being transported. The data, previously identified as the payload, begins somewhere in the payload envelope. The actual starting point will vary, as we will see later. The Path Overhead begins when the payload begins; because it is unique to the payload itself, it travels closely with the payload. The first byte of the payload is the first byte of the Path Overhead.

A word about nomenclature: two distinct terms, *Synchronous Transport Signal* (STS) and *Optical Carrier Level* (OC), are often used (incorrectly) interchangeably. They are used interchangeably because although an STS-1 and an OC-1 are both 51.84 Mbps signals, one is an electrically-framed signal (STS) whereas the other describes an optical signal (OC). Keep in mind that the signals SONET transports usually originate at an electrical source such as a T-1. This data must be collected and multiplexed at an electrical level before being handed over to the optical transport system. The optical networking part of the SONET system speaks in terms of OC.

Let's pause for a moment to consider the actual transmission of a SONET frame. Several years ago, while teaching a course on basic telecom-

Figure 2-5
A SONET frame showing the principal components.

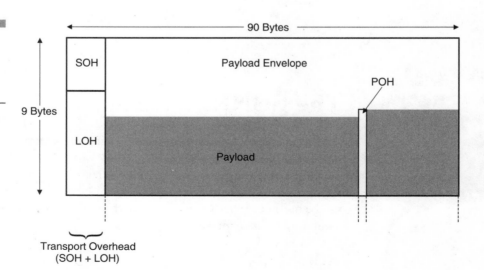

munications to a group of telecommunications lawyers, one of them asked me a very interesting question. While examining the 810-byte SONET frame shown in Figure 2-6, she asked, "how does SONET multiplex those nine rows to get them into the transmission facility?" It took me a moment to understand her question, but when I finally did I realized that the question deserved an answer because she was probably the first person to articulate a question that many before her had no doubt wondered about. The SONET frame is transmitted serially on a row-by-row basis, as shown in Figure 2-7. For purposes of clarity, I have simplified the frame by separating the rows and left-justifying them. The SONET multiplexer transmits (and therefore receives) the first byte of row one, all the way to the 90th byte of row one, then wraps to transmit the first byte of row two, all the way to the 90th byte of row two, and so on, until all 810 bytes have been transmitted. We draw the SONET frame structure as a 9-by-90-byte box, because if we were to draw it as a linear transmission stream, the frame would wrap completely around the classroom and require multiple pages at the bottom

Figure 2-6
A 90x90-byte SONET frame.

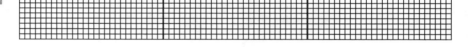

Figure 2-7
Serial transmission of data in SONET.

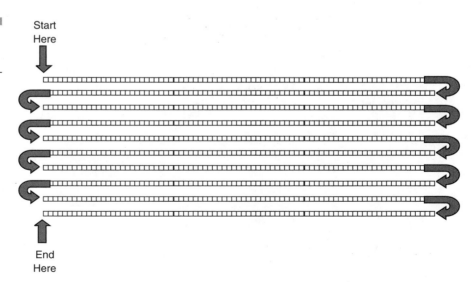

of this book, like a timeline in a children's book on dinosaurs (the Jurassic, the Triassic, the Plesiochronous, the Miocene, and so on).

Of course, the careful reader will realize that because the rows are transmitted serially, the many overhead bytes do not all appear at the beginning of the transmission of the frame; instead, they are peppered along the bit stream, like highway markers. For example, the first two bytes of overhead in the Section Overhead are the framing bytes, followed by the single-byte signal identifier. The next 87 bytes are user payload, followed by the next byte of Section Overhead; in other words, 87 bytes of user data are between the first three Section Overhead bytes and the next one! The designers of SONET were thinking clearly the day they came up with this because each byte of data appears just when it is needed. That is truly remarkable!

Also, notice the dotted lines descending from the bottom of the frame in Figure 2-5. This is to indicate one of the rather remarkable things about SONET. As we said earlier, because of the unique way that the user's data is mapped into the SONET frame, the data can actually start pretty much anywhere in the payload envelope. The payload is always the same number of bytes, which means that if it starts late in the payload envelope, it may well run into the payload envelope of the next frame! In fact, this happens more often than not, but it's OK—SONET is equipped to handle this odd behavior. We'll discuss this shortly.

SONET Bandwidth

The SONET frame consists of 810 eight-bit bytes, and like the T-1 frame, it is transmitted once every 125 μsec (8,000 frames per second). Doing the math, this works out to an overall bit rate of

810 bytes/frame/8 bits/byte/8,000 frames/second = 51.84 Mbps,

the fundamental transmission rate of the SONET STS-1 frame.

That's a lot of bandwidth—51.84 Mbps is slightly more than a 44.736 Mbps DS-3, a respectable carrier level by anyone's standard. What if more bandwidth is required, however? What if the user wants to transmit multiple DS-3s or perhaps a single signal that requires more than 51.84 Mbps, such as a 100 Mbps Fast Ethernet signal? Or for that matter, what about a payload that requires less than 51.84 Mbps? In those cases, we have to invoke more of SONET's magic.

The STS-N Frame

In situations where multiple STS-1s are required to transport multiple payloads, all of which fit in an STS-1's payload capacity (such as the multiple DS-3s shown in Figure 2-8), SONET enables the creation of what are called STS-N frames, where N represents the number of STS-1 frames that are multiplexed together to create the frame. If three STS-1s are combined, the result is an STS-3. In this case, the three STS-1s are brought into the multiplexer and *byte interleaved* to create an STS-3, as shown in Figure 2-9. In other words, the multiplexer selects the *first* byte of frame one, followed by the *first* byte of frame two, followed by the *first* byte of frame three. Then it selects the *second* byte of frame one, followed by the *second* byte of frame two, followed by the *second* byte of frame three, and so on, until it has built an interleaved frame that is now three times the size of an STS-1: 9×270 bytes instead of 9×90. Interestingly (and impressively), the STS-3 is still generated 8,000 times per second.

The technique described above is called a *single stage multiplexing process* because the incoming payload components are combined in a single step. A two-stage technique is also commonly used. For example, an STS-12 can be created in two ways. Twelve STS-1s can be combined in a single stage process to create the byte interleaved STS-12; alternatively, four groups of three STS-1s can be combined to form four STS-3s, which can then be further combined in a second stage to create a single STS-12. Obviously, two-stage multiplexing is more complex than its single-stage cousin, but both are used.

Figure 2-8
DS-3 transmission in SONET.

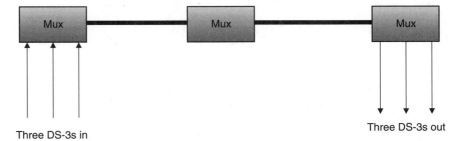

Three DS-3s in

Three DS-3s out

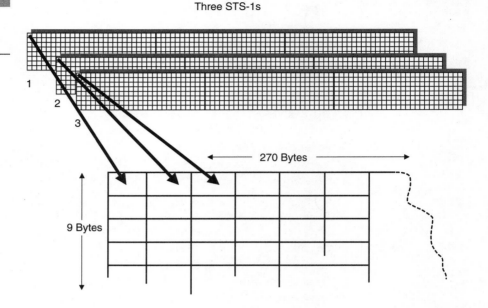

Figure 2-9
Byte interleaving in
SONET.

Three STS-1s

270 Bytes

9 Bytes

NOTE: *The overall bit rate of the STS-N system is N×STS-1. However,*
the maximum bandwidth that can be transported is STS-1, but N of them
can be transported! This is analogous to a channelized T-1.

The STS-Nc Frame

Let's go back to our Fast Ethernet example mentioned earlier. In this case,
51.84 Mbps is inadequate for our purposes because we have to transport
the 100 Mbps Ethernet signal. For this we need what is known as a *con-*
catenated signal. One thing you can say about SONET: it doesn't hurt for
polysyllabic vocabulary.

On the long, lonesome stretches of outback highway in Australia, unsus-
pecting car drivers often encounter a devilish vehicle known as a road train.

Figure 2-10
Transporting super-rate frames in SONET.

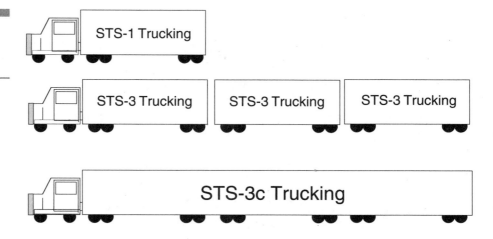

Imagine an eighteen wheel tractor-trailer (see top drawing, Figure 2-10, for a remarkable illustration) barreling down the highway at 80 miles per hour, but now imagine that it has six trailers—in effect, a 98-wheeler. These things give passing a whole new meaning. If a road train is rolling down the highway pulling three 50-foot trailers (middle drawing, Figure 2-10), then it has the ability to transport 150 feet of cargo, but only if the cargo is segmented into 50-foot chunks.

But what if the trucker wants to transport a 150-foot long item, such as a Chernobyl-enhanced blue whale or a typical Australian earthworm? In that case, a special trailer must be installed that provides room for the 150-foot payload (bottom drawing, Figure 2-10).

If you understand the difference between the second and third drawings, then you understand the difference between an STS-N and an STS-Nc. The word concatenate means "to string together," which is exactly what we do when we need to create what is known as a *super-rate frame*—in other words, a frame capable of transporting a payload that requires more bandwidth than an STS-1 can provide, such as our 100 Mbps Fast Ethernet frame. In the same way that an STS-N is analogous to a channelized T-1, an STS-Nc is analogous to an *unchannelized* T-1. In both cases, the customer is given the full bandwidth that the pipe provides; the difference lies in how the bandwidth is parceled out to the user.

Overhead Modifications in STS-Nc Frames

When we transport multiple STS-1s in an STS-N frame, we assume that they may arrive from different sources. As a result, each frame is inserted into the STS-N frame with its own unique set of overhead. When we create a concatenated frame, though, the data that will occupy the combined bandwidth of the frame derives from the same source. For example, if we pack a 100 Mbps Fast Ethernet signal into a 155.53 Mbps STS-3c frame, we only need to pack one signal. It's pretty obvious, then, that we don't need three sets of overhead to guide a single frame through the maze of the network[1]. For example, each frame has a set of bytes that keep track of the payload within the synchronous payload envelope. Because we only have one payload, we can eliminate two of them. The Path Overhead that is unique to the payload can similarly be reduced because a column of it is available for each of the three formerly individual frames. In the case of the pointer that tracks the floating payload, the first pointer continues to perform that function; the others are changed to a fixed binary value that is known to receiving devices as a *concatenation indication*. The details of these bytes will be covered later in the overhead section.

Transporting Sub-Rate Payloads: Virtual Tributaries

Let's now go back to our Australian road-train example. This time he is carrying individual cans of Fosters Beer. From what I remember about the last time I was dragged into an Aussie pub (and it isn't much), the driver could probably transport about six cans of Fosters per 50-foot trailer. So now we

---

[1] A significant chicken-and-egg problem arises in the flow of this book, and I apologize to the reader up front for it. In order to understand basic SONET frame structures and functions, the details of the SONET overhead should be discussed first. To understand the functions of the SONET overhead bytes, though, the basics of SONET frame structures and functions should be introduced first. I have chosen to do the frame structures and functions first, the result of which is that readers may find themselves flipping back and forth between the two sections. Sorry about that!

have a technique for carrying payloads smaller than the fundamental 50-foot payload size. This analogy works well for understanding SONET's ability to transport payloads that require less bandwidth than 51.84 Mbps, such as T-1 or traditional 10 Mbps Ethernet.

When a SONET frame is modified for the transport of sub-rate payloads, it is said to carry *virtual tributaries* (VTs). Simply put, the payload envelope is chopped into smaller pieces that can then be individually used for the transport of multiple lower-bandwidth signals.

Creating Virtual Tributaries

To create a virtual tributary-ready STS, the synchronous payload envelope is subdivided. An STS-1 comprises 90 columns of bytes, four of which are reserved for overhead functions (Section, Line, and Path). This leaves 86 for actual user payload. To create virtual tributaries, the payload capacity of the SPE is divided into seven, 12-column pieces called *virtual tributary groups*. Math majors will be quick to point out that $7 \times 12 = 84$, leaving two unassigned columns. These columns, shown in Figure 2-11, are indeed unassigned and are given the rather silly name of *fixed stuff.*

Now comes the fun part. Each of the VT groups can be further subdivided into one of four different VTs to carry a variety of payload types, as shown in Figure 2-12. A VT1.5, for example, can easily transport a 1.544 Mbps

Figure 2-11
Fixed stuff in SONET frames configured to carry virtual tributaries.

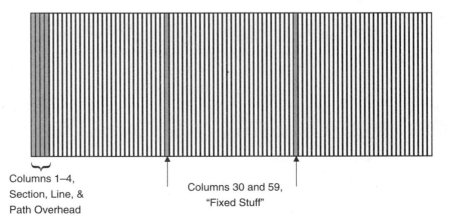

Columns 1–4,
Section, Line, &
Path Overhead

Columns 30 and 59,
"Fixed Stuff"

Figure 2-12
Virtual tributaries in
SONET.

VT Type	Columns/VT	Bytes/VT	VTs/Group	VTs/SPE	VT Bandwidth
VT1.5	3	27	4	28	1.728
VT2	4	36	3	21	2.304
VT3	6	54	2	14	3.456
VT6	12	108	1	7	6.912

signal within its 1.728 Mbps capacity, with a little room left over. A VT2, meanwhile, has enough capacity in its 2.304 Mbps structure to carry a 2.048 Mbps European E-1 signal, with a little room left over. A VT3 can transport a DS-1C signal, whereas a VT6 can easily accommodate a DS-2, again, each with a little room left over.

One aspect of virtual tributaries that must be mentioned is the mix-and-match nature of the payload. Within a single SPE, the seven VT groups can carry a variety of different VTs. However, each VT group can carry only one VT type.

That "little room left over" comment earlier is, by the way, one of the key points that SONET and SDH detractors point to when criticizing them as legacy technologies, claiming that in these times of growing competition and the universal drive for efficiency, they are inordinately wasteful of bandwidth, given that they were designed when the companies that delivered them were monopolies and less concerned about such things than they are now. We will discuss this issue in a later section of the book. For now, though, suffice it to say that one of the elegant aspects of SONET is its ability to accept essentially any form of data signal, map it into standardized positions within the SPE frame, and transport it efficiently and at a very high speed to a receiving device on the other side of town or the other side of the world.

Creating the Virtual Tributary Superframe

You will recall that when DS-1 frames are transmitted through modern networks today, they are typically formatted into extended superframes in order to eke additional ability out of the comparatively large percentage of overhead space that is available. When DS-1 or other signals are transported via an STS-1 formatted into virtual tributary groups, four consecutive STS-1s are ganged together to create a single VT Superframe, as shown in Figure 2-13. To identify the fact that the frames are behaving as a VT Superframe, certain overhead bytes are modified for the purpose, as discussed in the section that follows on SONET overhead.

SONET Overhead

Let's now look at the three types of SONET overhead in more detail. As we mentioned earlier, SONET overhead comes in three flavors: Section, Line, and Path.

The word *overhead* often has negative connotations because it usually refers to extraneous, unnecessary, poorly managed, or underutilized resources. For SONET, however, this is far from the case. In fact, the overhead that the SONET standard defines is perhaps the most important ability that the technology offers. The overhead bytes are well planned and

Figure 2-13
VT Superframe.

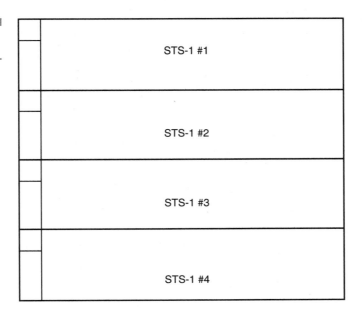

carefully thought-out, and as a result, offer network providers the ability to manage their resources with enormous granularity across the wide area.

Section Overhead

The Section Overhead comprises nine bytes, as shown in Figure 2-14. Their individual functions are described in the following section. Remember the reason for the Section Overhead: it contains information that every device in the network uses, from the originating device all the way to the destination device.

A1, A2: The A1 and A2 bytes provide a unique framing pattern at the beginning of the frame that is used to identify the beginning of the frame to receiving equipment for synchronization purposes. The pattern is the hexadecimal number 0xF628, which in binary is 1111 0110 0010 1000. When transmitting STS-N frames, the A1 and A2 bytes must appear in the overhead of each STS-1 in the STS-N frame.

J0/Z0: A unique sequence number called the STS-1 ID identifies every STS-1. It is carried in the J0/Z0 byte. Formerly known as the C1 byte, it is used to identify the originating office of a frame as it makes its way across the network. Because SONET transmission potentially involves multiple levels of interleaved multiplexing, the STS-1 ID provides a way to uniquely identify each frame of data.

When the C1 byte was first defined, SONET transmission speeds were relatively limited, with little thought that network design engineers would actually push the limits of the technology to the speeds that the standard currently provides. The format of the eight bits for identifying STS-1s then was to simply number the STS-1s of the STS-N (in hexadecimal) as follows:

Figure 2-14
SONET Section Overhead.

A1: Framing	A2: Framing	J0/Z0: STS-1 ID
B1: BIP-8	E1: Orderwire	F1: User
D1: Data Com	D2: Data Com	D3: Data Com

0000–0001 for the first STS-1 in the series, 0000-0010 for the second, 0000-0011 for the third, and so on. The problem is that the transmission levels that we currently operate at may well soon exceed this numbering scheme. The final standards for the use of the J0/Z0 byte are still under development; until they are complete, service providers use it the same way the C1 byte was originally intended to be used. However, under the new standard, it is clear that because all the STS-1s in an STS-N come from the same source, the STS-1 ID will only be defined in the first STS-1 of the STS-N. The remaining bytes in the succeeding frames will be used for anticipated growth and are designated as Z-bytes; these Z-bytes are designated for functions that have not yet been defined within the SONET parameters.

B1: The B1 byte is known as the *Bit Interleaved Parity* byte (BIP-8). It provides an eight-bit parity check that is used for error detection. In practice, the parity check is performed across all bytes of the prior frame and placed in the B1 byte of the current frame prior to scrambling.

A word about scrambling: in metallic (non-optical) digital transmission systems, zeroes and ones are often represented as zero voltage and either positive or negative voltage, respectively, as shown in Figure 2-15. This technique is known as *alternate mark inversion* (AMI). The technique was specifically chosen to ensure what is known as *ones density*.

The regenerators or repeaters that sprout along a digital transmission facility are not particularly intelligent devices. In fact, they are not much smarter than the wire they are connected to. They do, however, perform the rather sophisticated task of retiming and reframing a weakened noisy digital signal when it arrives; they do this by first finding the incoming framing information. What, though, gives a stupid device the ability to time and synchronize the inbound signal?

The answer is quite simple. The repeater looks for the digital pulses that represent ones in the bit stream, and because they arrive at random but

Figure 2-15
Alternate Mark
Inversion (AMI).

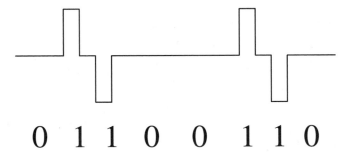

0 1 1 0 0 1 1 0

precise intervals, the repeater has the ability to derive its own timing from the incoming signal. What happens, however, if the incoming bit stream should comprise a long series of zeroes—which in digital parlance would be a flatline? If that were to happen, the repeater would lose its ability to derive a synchronization signal from the data stream and errors could occur.

To prevent this from happening, international transmission standards place certain stringent standards on the devices that originate bit streams. One of these standards, used in both SONET and SDH, dictates that all data carried in a frame will be scrambled prior to being transmitted, with the exception of the initial framing bytes. All other bytes pass through a processor that manipulates the data (for the bit-weenies in the audience, it uses the polynomial $1+x^6+x^7$), changing it to ensure that an appropriate "richness of ones" occurs in the bitstream, avoiding the flatline problem. At the far end of the span, a similar process unscrambles the data, returning it to its original value. The initial framing bytes are not scrambled because they contain the unique framing code that is used to identify the beginning of the frame.

The B1 byte is only defined in the first STS-1 of an STS-N signal.

E1: The Orderwire byte is a 64-Kbps voice channel that can be used by technicians to communicate while troubleshooting between repeater spans. When building STS-N frames, the E1 byte is only defined in the first STS-1 of the STS-N for obvious reasons. According to an informal survey conducted by the author, relatively few companies actually use the E1 byte. In order for it to work, an *Electrical Order Wire* (EOW) interface is required, usually a small rack-mounted device that interfaces between a traditional copper-seeking device such as a telephone or butt-in and the SONET E1 channel.

F1: The F1 byte, called the user byte, is user configurable and can be employed for a variety of purposes. It is reserved for the use of the service provider as a transport byte for network management-application information, although many service providers today use it to transport maintenance communications and configuration data. Because the byte's specific use is not specified in a standard, it is extremely flexible, but lends very little to the overall goal of interoperability.

D1, D2, D3: The *Data Communications Channel* (DCC) bytes provide a 192-Kbps communications channel between section-terminating devices and are used to transport operations, administration, maintenance, and provisioning (OAM&P) information such as control signals, monitoring, alarm information, and so on. Although the DCC bytes are fully in alignment with the nascent *Telecommunications Management Network* (TMN) standards that continue to evolve, it will probably be some time before their use is clearly defined among operators to ensure interoperability.

Line Overhead

The Line Overhead, shown in Figure 2-16, occupies the lower 18 bytes of the Transport Overhead (the first three columns of the STS-1). It is not used by the regenerators along the span as the Section Overhead is, but it is accessed by all other devices along the span. Compared to the functions found in the Section Overhead, the Line Overhead is significantly more complex.

H1 / H2: pointer bytes: To understand the role of the pointer bytes, it is first useful to talk about the international shipping business. Most people have driven past, flown over, or visited one of the world's major ports, where ships laden with international shipping containers (those big metal SeaLand boxes) arrive and tie up under those massive white cranes that look for all the world like the Imperial Walkers from George Lucas' Star Wars series, shown in Figure 2-17. The cranes remove the containers and load them onto trucks or trains that pass below. If everything happens as it should, a good and careful crane operator can load the train quite effectively.

Figure 2-16
SONET Line
Overhead.

H1: Pointer	H2: Pointer	H3: Pointer Action
B2: BIP-8	K1: APS	K2: APS
D4: Data Com	D5: Data Com	D6: Data Com
D1: Data Com	D2: Data Com	D3: Data Com
D10: Data Com	D11: Data Com	D12: Data Com
S1/Z1:Sync Status/Growth	M0 or M1/Z2 REI-L/Growth	E2: Orderwire

Figure 2-17
Cranes at a
commercial port.

Now imagine the following scenario. A number of ships are tied up at the dock behind the crane, and their deck crews are each unloading the containers and placing them on a belt that takes them to the top of the crane to be loaded onto a train that is passing below, as shown in Figure 2-18. The train, incidentally, is an endless train of flatbed cars that passes below the crane at a never-changing speed. The crane operator's job is to pick up an arriving container, position it over the train, and place it on a passing train car. As we said earlier, if all goes well, the operator should be able to place a container squarely on each car, ensuring that no train cars are empty. That, after all, would be wasteful.

Of course, a number of variables could affect the efficacy with which the operator does his or her job. For example, it is most certainly the case that the deckhands unloading the ships operate at different speeds and with different degrees of efficiency. Some will do so extremely quickly, whereas others will be less organized. The result is that in some cases the containers will arrive at the top of the crane at a regular pace, giving the operator a steady stream of payload to work with. Alternatively, they may arrive sporadically, which could lead to the occasional empty train car. And of course, some days the crews may want to get off the ship as soon as possible, which means that containers could arrive at the top of the crane faster than the operator can handle them. In that case, the crane operator would need a place to put the excess containers until the pace abates enough to handle them.

The operator, then, has two techniques available to him or her to load the train. The first is to carefully match the loading rate of the arriving containers to the unchanging speed of the train, assuming that the container arrival rate is adequate to allow this technique, which in the end will

Figure 2-18
Loading the crane.

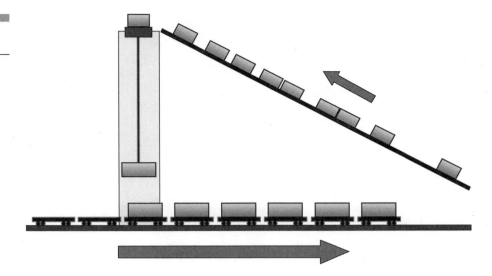

ensure that no flatbed will ever be empty. The problem still exists, of course, that traffic could arrive too fast, in which case some kind of parking area is needed. This is clearly complicated, requires a very talented operator, and assumes that the loading crews work very closely in order to synchronize their efforts to ensure maximum port loading and unloading operational efficiency.

The second technique is to simply not worry about it. In fact, as long as the operator knows that the crane is positioned correctly above the train as it passes underneath, he or she doesn't even have to look. As a container arrives, the operator simply drops it, and it lands on the train wherever it lands with little regard for alignment on a train car-by-train car basis, as shown in Figure 2-19. In this case, a lot of space on the train is wasted, but the upside is that the port assumes that train car space is cheap and that they save money because they don't have to worry about overly skilled (and therefore expensive) stevedores or crane operators. If the containers arrive efficiently, great; if not, well, no big deal.

Of course, it is potentially a big deal for the customer at the other end of the train line who is looking for his or her payload. The customer has been told that the shipment is on train car number 576. A well-run train system would clearly load the containers squarely on the cars, yet when the car arrives at the customer's location and he or she looks at the head of the car, as shown in Figure 2-20, all the customer sees is empty space. The

Figure 2-19
Wasted space when loading the crane.

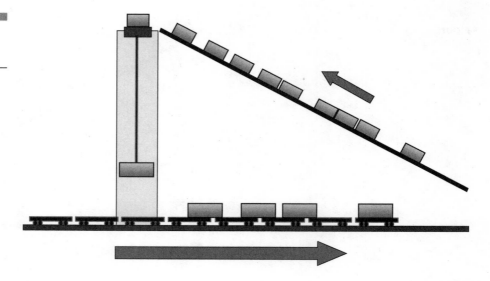

Figure 2-20
Finding the payload
n the SONET train.

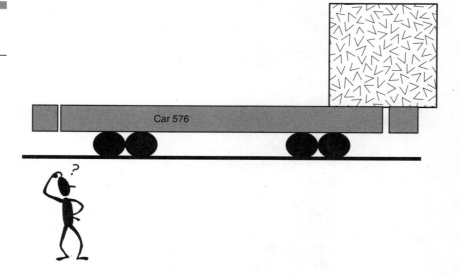

Car 576

customer's cargo is there, of course: it's just 37 feet to the right. Because the customer expects it to be aligned with the train car, he or she doesn't think to look over there.

What if we had a slightly different system, as shown in Figure 2-21? In this case, we tell the customer that his or her cargo may not begin right at the head of the train car. What will be at the head of the train car, however, is a bill of lading that tells the customer exactly where he or she has to look to see the beginning of the shipment. That way the complexity of guaranteeing the exact location of the container is eliminated. Furthermore, if the container should shift back and forth slightly on the train car as it makes its way down the track, no problem: every time the car goes through a switching yard, a conductor checks its location and updates the information in the bill of lading before the car leaves the yard.

If you understand this analogy, then you understand how SONET uses payload pointers, the first two bytes in the Line Overhead.

H1 and H2 are used to measure the offset, or distance, that exists between the pointer and the beginning of the payload. The reason for its existence is relatively simple: because of slight differences in timing signals between the devices that a SONET signal passes through as it makes its way across the network, or because of the slight signal phase variation (known as *jitter)* that can take place, the actual starting point of the signal can vary by as much as a byte forward and backward. The payload pointer has the ability to adjust itself according to the relative position of the payload, thus giving a receiving device the ability to actually find the payload in a received frame. How it does this is rather interesting.

Figure 2-21

Fixing the problem with a manifest pointer!

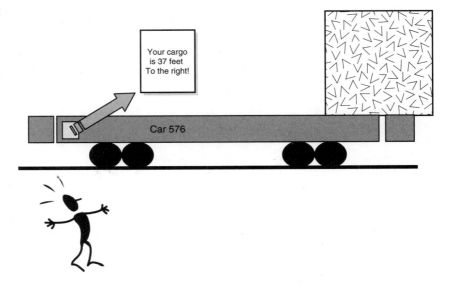

The 16 bits of the payload pointer are divided into three functional groupings. The first four bits of the H1 byte are called the *New Data Flag* (NDF). When a payload is constructed (and a new pointer value is introduced), the initial value of the four bits is 0110, as shown in Figure 2-22. Should the pointer value change due to an adjustment in the position of the payload, the NDF is inverted (0110 becomes 1001). This inversion indicates to a receiving device that the payload has shifted, and to stand by for instructions, which follow. The remaining 10 bits, some of which are the actual pointer, stay the same, always pointing to the first byte of the Path Overhead and therefore to the beginning of the payload.

The next two bits are always set to zero, and serve as nothing more than place keepers in the overall structure of the pointer.

The remaining 10 bits, which straddle the H0 and H1 bytes, are the actual payload pointer. These bytes take on a specific binary value between zero and 783 (the size of the STS frame, 810, minus the 27 bytes of Transport Overhead). For example, if the pointer value is 0100001011, the binary equivalent of decimal number 267, the first byte of the Path Overhead (and therefore the payload) begins exactly 267 bytes after the H3 byte, which will be discussed shortly. Thus, a receiving SONET device only has to be able to stomp its foot like a horse and count in order to find the payload it may be looking for.

Figure 2-22
SONET payload
pointer.

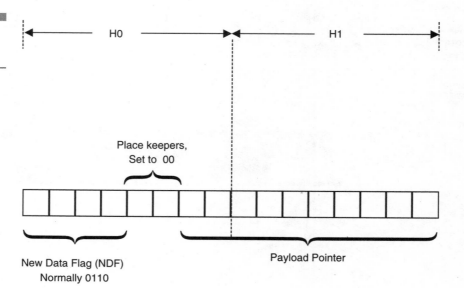

Of course, it would be too easy to make the pointer's behavior so simple. The 10 bits of the actual pointer are divided into five *increment bits* and five *decrement bits.* The odd bits are the increment bits; the even bits are the decrement bits. Whenever a pointer adjustment takes place, the absolute value of the pointer changes dramatically. In the event of a negative adjustment of the pointer, the decrement bits are inverted, telling the receiver that the adjustment has occurred (in concert with the inversion of the New Data Flag). In the event of a positive adjustment, the increment bits are flipped.

This should sound a little strange to the reader. After all, this results in the conversion of the pointer value from a binary number that indicated the absolute location of the beginning of the SPE to a nonsensical number that now indicates nothing about the actual location of the payload. Remember, though, that the accuracy of SONET is such that the pointer value will never need to wander more than a single byte in either direction. Thus, it doesn't matter that the value of the pointer changes drastically; if the receiver sees the NDF and then notices that *at least three of the five pointer bytes changed,* then it knows that either a positive or negative justification took place and that it should modify its search accordingly.

We don't know about you, but we believe that somebody was practicing better life through chemistry when they thought this stuff up. It is truly amazing.

H3: *pointer action byte*: The pointer action byte is the byte that actually compensates for the movement of the SPE within the STS frame. As we saw in our loading dock example, the payload may actually arrive faster than the train cars do or may not arrive in time to align the payload with a passing train car. In SONET systems, the pointer action byte compensates for this. In the event that the payload rate exceeds the frame capacity (in other words, more than 783 bytes are ready to be transmitted within a single 125 _s period), a single excess byte can be carried in the pointer action byte position, thus expanding the SPE from 783 bytes to 784. This technique is called *negative timing justification.* As the reader might expect, an opposite technique is available called *positive timing justification*, which is used whenever the SPE is short a byte—in which case the H3 byte pushes itself forward in the frame, providing the missing byte to the system as it attempts to build a 783-byte payload.

One thing that should be noted: SONET timing systems are extremely accurate and thus only allow the payload to shift by as much as a single byte in either direction. However, this technique, known in SONET parlance as *floating mode,* is complex and expensive, and many service providers do all they can to avoid it because it requires the deployment of sophisticated pointer processing. The alternative to floating mode is called *locked mode,* in which case the payload pointer points to the first available payload byte in the next frame, which is row one, column four, as shown in Figure 2-23. However, this technique also has its downside; it requires buffering ability and complex network-management techniques, and it assumes that network-wide timing sources are operating in synchronous lock-step—an assumption that may not be valid in today's network environment, as stable as it is. So both techniques are employed and will continue to be until network-wide timing becomes a reality. Timing will be discussed in more detail in a later section of the book.

A final note about the H3 byte: it represents an integral part of the payload pointer function and therefore, like the pointer itself, must be provided for every STS within an STS-N frame. However, unless it carries justification information, it is ignored.

B2: BIP-8: Sound familiar? B2 is another bit-interleaved parity byte and is used to carry error-checking information. It calculates its value based on the Line Overhead (it does not include the Section Overhead) and payload of the previous frame before it is scrambled for transmission. The information is then placed in the current frame before it is scrambled. The BIP-8 is required for all STS in an STS-N.

K1/K2: automatic protection switching: One of the principal advantages of SONET is its ability to detect a failure in the network and switch traffic to a backup transmission span automatically if one is available, ensuring protection of customer data. This technique is called *automatic protection switching* (APS). If the network is deployed across a ring architecture, the K1 and K2 bytes are used to switch from one ring to another. Consider, for example, the ring shown in Figure 2-24. Although many companies use four-fiber rings, for our purposes this one has two. The outer ring, in this

Figure 2-23
Locked mode in
SONET.

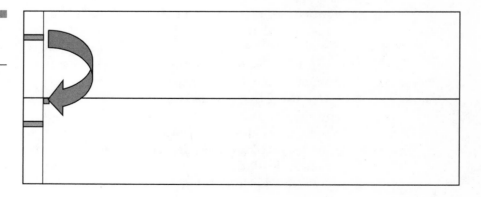

case the active ring that transports customer traffic, flows in a clockwise direction, which means that data transmitted from multiplexer A to multiplexer B would have a single-hop path to get there. The inner ring, which flows in a counter-clockwise direction, is the protect or backup path; it carries no user data, but monitors the condition of the network by paying close attention to alarm data carried in the APS bytes.

Let's assume that a terrorist backhoe driver, shown in Figure 2-25, craftily cuts the active fiber between multiplexers B and C as shown in Figure 2-26. When this occurs, a number of rather interesting events take place. The first thing that happens is that in short order, multiplexer C detects a loss of signal. Please note that multiplexer C has no idea what caused the loss, it just knows that the signal's gone. It immediately goes into recover mode. It first transmits an *alarm indication signal* (AIS) to its next downstream neighbor, multiplexer D, which will in turn pass it on to the next device in the ring.

Next, multiplexer C sends an upstream *far-end receive failure* (FERF) notification on the backup span to multiplexer B, telling the upstream device that it is no longer receiving a signal from it for some reason. Next, it initiates automatic-protection switching. Within 50 ms, the ring switches traffic from the primary ring to the protect ring, thus ensuring the integrity of transmitted traffic. This technique has one downside; whereas traffic

Figure 2-24

Dual counter-rotating ring in SONET.

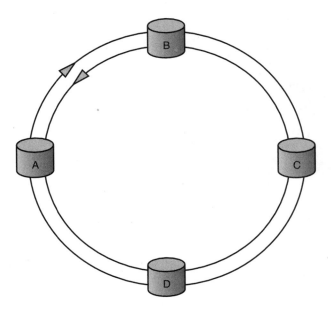

Figure 2-25
The dreaded backhoe.

Figure 2-26
Single fiber cut.

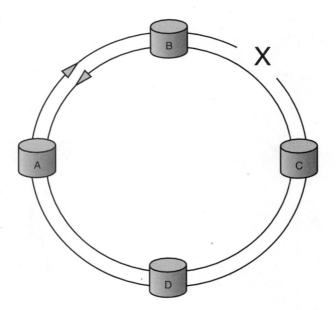

from multiplexer B to multiplexer C was a single hop, now it must endure three hops, leading to an increase in overall delay, which could be problematic for some applications.

Of course, the chance always exists that the backhoe driver is truly diligent or that the two fibers are diversely routed on either side of the same conduit, and that as a result, both fibers are cut between B and C, as shown in Figure 2-27. In this case, something that borders on magic happens: the ring heals, as illustrated in Figure 2-28. Because of the APS monitoring data that is transported around the ring as part of the SONET overhead, B and C soon realize that they no longer have continuity between them in either the upstream or downstream direction, so they wrap the two rings internally, resulting in the creation of a single ring shaped like a gnocchi. As a result, not even the destruction of both fibers stops transmission around the ring. Survivability is clearly one of the greatest advantages that SONET brings to the network domain, and one of the reasons that its predicted imminent demise may not be so imminent.

The APS process is enormously complex, and its capabilities and inner workings are only touched on here. For additional information, readers should consult *ANSI T1.105.01-1998*, an impressive document that covers the roles and responsibilities of the 16 APS bits.

Figure 2-27
Dual fiber cut.

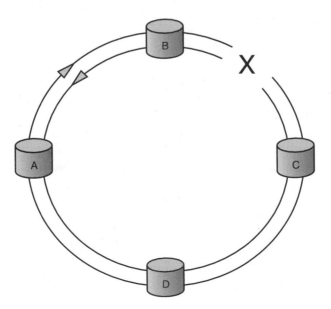

Figure 2-28
Failure causes ring to
wrap.

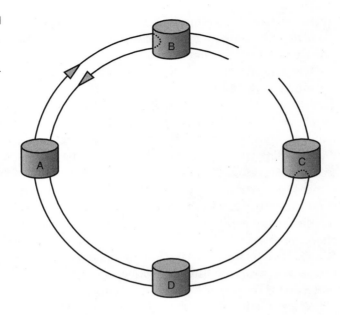

D4-D12: *Data Com Channels* (DCC): These nine bytes comprise a 576-Kbps message channel that SONET network elements can use to transport network management (OAM&P) messages. The information carried in these bytes is similar to that carried in the Section Layer's Data Com channels, but in this case is specific to the Line Layer. These bytes are only defined for the first STS in an STS-N. Like the Section Overhead's DCC bytes, they are designed to use TMN protocols, although many service providers rely on TL1 or other solutions pending the widespread deployment of TMN.

S1/Z1: *synchronization status/growth*: This byte is still in development in terms of its specific responsibilities. Of its eight bits, the last four (bits five through eight) have been reserved for the transport of timing information that permits a network element to choose a specific clock source based on its own selection parameters as a way to prevent timing problems within the network. This byte is defined in the first STS of an STS-N only; all remaining S1 bytes are designated as Z (growth) bytes.

M0/M1/Z2/REI-L: This particular byte is something of a chameleon in that it can take on a number of roles depending on the nature of the payload with which it is associated. If the payload being transported is a single STS-1, then the byte becomes the M0 byte and is used for error control;

then, it is called the *Remote Error Indication-Line Level* (REI-L). In earlier versions of SONET, one of the signals that was closely monitored was called a *Far-End Block Error* (FEBE), which indicated the presence of bit-level errors. It has been replaced by the REI-L, and its job is to transport the BIP-8 information back to the presumed source of the problem. The last four bits of the M0 byte are used for this function; the first four are undefined.

If the payload being transported is an STS-N rather than a single STS, then the byte becomes the M1 byte, which is only defined in the third STS of an STS-N frame (This makes sense because the next level up from an STS-1 is an STS-3). Its responsibilities are essentially identical to that of the M0 byte. If the byte is not used for M0 or M1 functionality, it is designated as a growth byte (Z2).

E2: *Orderwire*: The E2 byte is functionally identical to the E1 byte discussed in the Section Overhead section (Is that like the Department of Redundancy Department?). It provides a standard 64-Kbps voice channel for maintenance communications between technicians and is only defined in the first STS of an STS-N.

That concludes the discussion of the Line Overhead, and therefore the discussion of the Transport Overhead (Section plus Line). We now move on to the final overhead component, the Path Overhead.

Path Overhead

The Path Overhead, you will recall, is an integral part of the SPE and is therefore processed wherever the payload is processed—specifically, at the end devices (usually customer CPE). The Path Overhead is a single column of nine bytes, and the first byte of the Path Overhead is the first byte of the payload, as shown in Figure 2-29.

The Path Overhead bytes are shown in Figure 2-30. Their functions are discussed below.

J1: *trace*: The path trace byte is used by a receiving device (usually referred to as a *path terminating device*) to ensure that it is indeed connected to the proper transmitting device. The field is used to generate a 64-byte repeating string that identifies the source of the received signal. The field can be programmed by the user any way they like and often contains an IP address, a *Common Language Location Code* (CLLC, or *click code*), a *Common Language Location Indicator* (CLLI, or *silly code*), or a telephone number that identifies the transmitting device. Relatively few end users actually populate this field; if it is used at all, the service provider usually populates it. If for some reason neither party uses the field, it is filled with null characters.

Figure 2-29
Location of Path
Overhead.

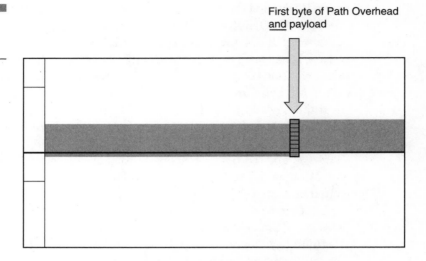

First byte of Path Overhead
<u>and</u> payload

Figure 2-30
Path Overhead bytes.

J1: Trace
B3: BIP-8
C2: Signal Label
G1: Path Status
F2: User Channel
H4: Indicator
Z3: Growth
Z4: Growth
Z5: Tandem Connection

B3: *Path BIP-8*: Like the BIP-8 that we saw in the Section and Line Overhead, the Path BIP-8 is used to error-check the content of the previous SPE before being scrambled. As with the Line Overhead (in reference to the Section Overhead), the Path BIP-8 does not include the Line or Section Overhead in its calculation.

C2: *STS path signal label*: The signal label byte is used to tell a receiving device what is actually contained in the SPE that is arriving in terms of the actual construction of the payload. This permits the simultaneous transport of multiple traffic types.

When SONET was first released, the transport of multiple traffic types was not an issue of grandiose proportions. After all, voice was almost universally the only traffic type transported in early versions of the standard, so many manufacturers *hard coded* the C2 byte to a binary value of 02, the standard coding for virtual tributaries (see Figure 2-31). As the standard matured and end user devices became sophisticated enough to accept and package multiple traffic types, the need to label the content of individual SPEs became more important. Many of the manufacturers had to go back and offer fixes to overcome the limitation of a single payload mapping value.

G1: *path status*: As it's name implies, the path status byte is used to communicate the overall transmission status of the duplex circuit to the originating device. In practice, it transports two indicators: the *Path Remote Error Indicator* (REI-P) in bits one through four, and the *Path*

Figure 2-31
Signal label mapping.

Value	S PE Content
00	Unequipped—not used for live data
01	Equipped—non-specific payload
02	Virtual tributaries
03	Virtual tributaries in locked mode
04	Asynchronous DS3 mapping
12	Asynchronous DS-4NA mapping
13	ATM cells
14	DQDB cells
15	Asynchronous FDDI
16	HDLC frames over SONET (used for IP transport)
CF	Experimental value for IP transported in PPP frames
FE	Test signal mapping, per ITU Recommendation G.707

Remote Defect Indicator (RDI-P) in bits five through seven, leaving a single unused bit.

F2: *path user channel*: The network service provider can use this byte to transport communications information such as network management data.

H4: *indicator*: The H4 byte is used to indicate the manner in which payload is mapped into an SPE, such as when the SPE is subdivided into virtual tributaries. At one time, the H4 byte was used as a way to indicate the boundaries between cells, but this mapping is no longer supported. One use that is still supported, even though the technology has fallen out of favor, is to carry link status information for *distributed-queue dual-bus* (DQDB) networks.

Z3–5: *growth*: Amazing how much DQDB continues to rear its head. These growth bytes can be used to transport DQDB layer-management data, but otherwise are reserved for future, as yet undetermined, applications.

Z5: *tandem connection*: My favorite telecommunications joke states, "the best thing about standards is that there so many to choose from." The *American National Standards Institute* (ANSI) has stipulated that the Z5 byte can be used as a maintenance channel for the transport of information between tandem switches, as well as a data communications channel (DCC) for path level management information. Telcordia (formerly Bellcore) still defines Z5 as a growth byte.

That concludes our discussion of the SONET overhead. It's a far cry from the utility of a frame bit in a T-1 frame or the reuse made of the 24 frame bits in an extended superframe, wouldn't you agree? It provides for the delivery of an enormous amount of ability and once again points to the longevity of the standard in spite of claims to the contrary.

One area that has not yet been discussed in this section is network timing.

Synchronization and Timing in SONET Networks

When the telephone company first began to deploy networks, they realized that they had a significant problem to overcome. The networks they deployed had a clear hierarchy to them, as shown in Figure 2-32. The box at the top labeled "1" represents the highest level of the switching hierarchy in a region (for example, North America). In this model, it provides a synchronization and timing signal to the devices immediately below it, labeled "2". These switches, in turn, provide timing to the third tier, and so on down the line.

The highest level of the network hierarchy required the most accurate timing signal of all because they were responsible for providing the reference-timing signal for all the subtending layers in the network. As a result, at the top of the network heap lived an extremely accurate cesium clock that beat out a standard-timing signal called the *Bell System Reference Frequency* (BSRF) that all tier-one network devices used as their timing source. These tier-one devices would in turn provide a timing signal to their downstream partners, which would do the same for theirs, and so on.

The only problem with this model is that all devices in the network are required to derive timing from a sole source, which involves enormous complexity when one considers the magnitude of this problem. Consider how many central office switches are available, not to mention digital cross-connect systems, carrier devices, and other components that require synchronization.

So why didn't the offices simply generate their own timing signals? The answer is quite simple. The first digital-transmission systems came about in the early 1960s and at that time, the state of the art in power supplies and timing circuitry was not good enough to guarantee the stability required to time a device as critically dependent on stable clocking as a high-end switch or multiplexer. The clocks would drift, which would in turn lead to timing slips, and the ultimate result was errors in the traffic being transported across the network. The only solution was to rely on a single, highly reliable (and enormously expensive) clock that could provide the necessary stability.

Figure 2-32
Timing hierarchy in
early networks.

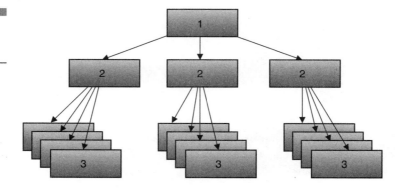

As time went on, power supplies became more stable, clock sources became more accurate, and alternatives to the BSRF became available. The most commonly used clock signal today derives from the *Global Positioning System* (GPS) satellites, and most central offices in North America have the ability to derive a GPS timing signal using a rooftop receiver (Figure 2-33) and feed the signal to a device known as a *Building Integrated Timing Supply*, or BITS clock, shown in Figure 2-34. This clock is accurate enough to provide timing to all devices in the office, and serves as the *Primary Reference Clock* (PRC). An office may actually have more than one feed from an accurate timing source to ensure redundancy, but it will only use one of the sources at a time as a timing signal for the devices in that office.

In data networks, lower-level devices routinely transmit traffic between each other, meaning that the overall system must ensure that two or more communicating devices in a transmission system be synchronized relative to one another to within certain tolerances. If we consider the example of a network of multiplexers that are communicating with each other, we see that the ability to synchronize the transmitted signal from a multiplexer is relatively simple. After all, the mux is generating the voltage pulses that indicate the mix of ones and zeroes to the receiver. How, though, can a multiplexer ensure that the received signal, that is, the signal arriving from another multiplexer, is synchronized as well? After all, the signal is coming

Figure 2-33
GPS receiver on roof of CO.

Figure 2-34
BITS clock in CO.

from another multiplexer, potentially in another network, synchronized by a different primary clock reference, which means that some drift or jitter may occur between the two.

The answer is actually quite simple. Multiplexers use a technique called *loop timing* or *phase locked loop timing* to ensure transmit and receive harmony. In loop timing, shown in Figure 2-35, each multiplexer uses its transmit clock's signal to provide timing to the received signal, telling the receiving circuitry when to look at the incoming bit stream. This ensures that although the muxes may be slightly out of whack with each other from a timing perspective, they will never be too far out of whack, thus guaranteeing that each receiving device in the network will be able to interpret the incoming information properly.

Today, central-office timing models are still hierarchical. They rely on a primary-reference clock signal, but instead of a single-reference clock providing a timing source for all devices in North America, each office has a primary clock, as shown in Figure 2-36. Thus, the GPS signal provides guidance to the PRC, which in turn provides timing and synchronization accuracy to the higher-tier toll switches in the network. The toll switches provide a signal to the local switches and digital cross-connect systems,

Figure 2-35
Loop timing.

Figure 2-36
GPS timing in
modern network.

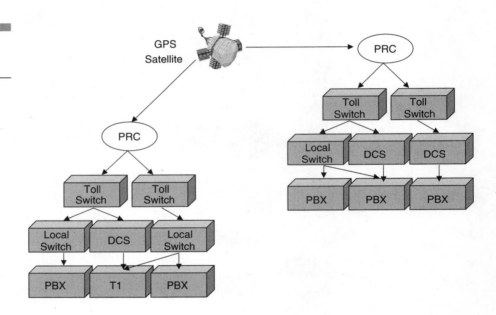

which time the PBXs and lower-tier multiplexers. Thus, all devices in the office derive their timing from a single source. The PRC is considered to be a Stratum 1 clock, whereas the toll switches are Stratum 2, the local switches and DCSs are Stratum 3, and the PBXs and T1 multiplexers are Stratum 1. As one descends in the hierarchy, the stringency of absolute accuracy relaxes. This is not to say that the clocks are inaccurate—as Einstein might have said, "Everything is relative, my child." A Stratum 4 clock, for example, has no real requirements for accuracy; the clocks found in PCs are Stratum 4 and they wander all over the place in terms of accuracy. Stratum 3 clocks are a little bit better; standards require that they have less than 255 timing slips in the first 24-hour period following a loss of clock reference. Stratum 2, on the other hand, must have less than 255 slips in the first 86 days following a loss of synchronization signal. I'd call that more accurate, wouldn't you? Finally, a Stratum 1 clock, used as the primary source in an office, is typically a cesium or rubidium atomic clock that is so accurate that it will lose a single second of accuracy approximately once every 250,000 years—good enough.

Because of the speed at which they operate, all SONET networks require timing signals that are Stratum 3 or better. Otherwise, pointer justification and other functions that were discussed earlier will not take place with appropriate levels of accuracy.

SONET Synchronization

SONET relies on a timing scheme called *plesiochronous timing*. As I implied earlier, the word sounds like one of the geological periods that we all learned in geology classes (Jurassic, Triassic, Plesiochronous, Plasticene, and so on). Plesiochronous derives from Greek and means "almost timed." Other words that are commonly tossed about in this industry are *asynchronous* (not timed), *isochronous* (constant delay in the timing), and *synchronous* (timed). SONET is plesiochronous in spite of its name (*Synchronous* Optical Network) because the communicating devices in the network rely on multiple timing sources and are therefore allowed to drift slightly relative to each other. This is fine because SONET has the ability to handle this with its pointer-adjustment capabilities.

The devices in a SONET network have the luxury of choosing from any of five timing schemes to ensure accuracy of the network. As long as the schemes have Stratum 3 accuracy or better, they are perfectly acceptable timing sources. The five are discussed below.

- *Line timing*: Devices in the network derive their timing signal from the arriving input signal from another SONET device. For example, an add-drop multiplexer that sits out on a customer's premises derives its synchronization pulse from the incoming bit stream and might provide further timing to a piece of CPE that is out beyond the ADM.

- *Loop timing*: Loop timing is somewhat similar to line timing; in loop timing, the device at the end of the loop is most likely a terminal multiplexer.

- *External timing*: The device has the luxury of deriving its timing signal directly from a Stratum 1 clock source.

- *Through timing*:   Similar to line timing, a device that is through timed receives its synchronization signal from the incoming bit stream, but then forwards that timing signal to other devices in the network. The timing signal then passes through the intermediate device.

- *Free running:*   In free-running timing systems, the SONET equipment in question does not have access to an external timing signal and must derive its timing from internal sources only.

One final point about SONET should be made. When the standard is deployed over ring topologies, two timing techniques are used. Either external timing sources are depended upon to time network elements, or one device on the ring is internally timed (free running), whereas all the others are through-timed.

In Summary

SONET is a complex and highly capable standard designed to provide high-bandwidth transport for legacy and new protocol types alike. The overhead that it provisions has the ability to deliver a remarkable collection of network management, monitoring, and transport granularity.

Some people believe that SONET is getting a little "long in the tooth." Current estimates, however, indicate that the market for SONET equipment will grow more than 25 percent per year for the next five years—a number that doesn't seem to indicate the death throes of a technological dinosaur. Without question, SONET is inefficient, wasteful, and inflexible. However, the services that it was designed to transport have no quarrel with its slothful ways, and most service providers have figured out clever and innovative ways to overcome the wasteful nature of the technology. As we progress, SONET will continue to provide high-bandwidth transport for some time to come and will advance in lockstep with the demands of the market. Will it eventually be relegated to the bone yard of technologies that have outlived their usefulness? Of course. However, that day is well over the horizon as of yet.

The interesting thing is that the Synchronous Digital Hierarchy (SDH) shares many of the same characteristics, as we will see in the next section. SONET, you will recall, is a limited North American standard, for the most part. The rest of the world awaits.

SDH Basics

The prior section of the book was devoted to SONET; in this section we examine the same issues as they relate to the *Synchronous Digital Hierarchy* (SDH). Many industry papers, books, and references refer to SDH as the "international version of SONET;" I prefer to view SONET as the North American version of SDH because far more countries rely on the SDH standard than on the SONET standard. That, however, is nothing more than a matter of preference and semantics, so we'll leave it at that.

A Bit of History

You will recall from earlier chapters that MCI's Bill McGowan played a major role in the development of the SONET standards. His efforts were also centrally important to the development of SDH.

When McGowan went before the *Interexchange Carrier Compatibility Forum* (ICCF) to request a standard for mid-span meet following divestiture, the ICCF turned around and sent the petition on to the ANSI T1 Committee. ANSI, founded in 1918, was one of the original standards bodies in the U.S., yet it is unique in that it issues no standards itself; instead, it "coordinates, facilitates, and approves" standards created by other bodies, most of them in the private sector. ANSI also represents the U.S. at the *International Organization for Standardization* (ISO).

ANSI comprises more than 300 individual standards bodies as well as a number of industry development groups. Among others, the ANSI organization staffs a number of subcommittees, one group of which falls under the T1 hierarchy. The T1 subcommittees manage ongoing research in telecommunications and computer technology. They include the T1E1 subcommittee, which is responsible for installation interfaces between carrier and customer equipment; T1M1, which is responsible for *operations, administration, maintenance, and provisioning* (OAM&P) standards; T1S1, which is responsible for signaling, network architecture, and services definition; T1Q1, which is responsible for performance; T1Y1, which governs standards related to specialized services such as video and data transport; and finally T1X1, which is responsible for timing, the definition of digital hierarchies, and network synchronization. As you might expect, both T1M1 (OAM&P) and T1X1 (synchronization and digital hierarchy) played special roles in the development of SONET and SDH standards. T1M1 coordinated the effort that guided the development of OAM&P standards for SONET, whereas T1X1 helped create standard SONET rates and the format of the transmission stream.

ANSI has always worked closely with a variety of other standards bodies including the *Alliance for Telecommunications Industry Standards* (ATIS), Bellcore (now Telcordia), the *International Telecommunications Union* (ITU), the *Institute of Electrical and Electronics Engineers* (IEEE), the *Electronics Industries Association* (EIA), the *European Computer Manufacturers Association* (ECMA), the *European Telecommunications Standards Institute* (ETSI, formerly CEPT), and most recently, the *SONET Interoperability Forum* (SIF). In the mid-1980s, Bellcore approached T1X1 with a solution to the problem of fiber transmission standardization. Prior to the proposal, any number of transmission systems were sold by any number of vendors, but they were proprietary. Bellcore's proposal suggested a methodology that would permit different vendors' optical devices and multiplexers to interoperate. They also proposed a bit-interleaved (not byte-interleaved) multiplexing standard, later known as SYNTRAN, which together with the interoperability model would lead to the rudiments of SONET.

A well-known model in the telecommunications industry is "The Apocalypse of the Two Elephants," illustrated in Figure 3-1. It describes the problem that faced would-be vendors of SONET equipment in the early days of its creation. Standards bodies always do an admirable job of trying to take into account every vendor's perspective on the direction that a standard should take during its development. In an ideal world, technology companies perform research, collect input from a large number of vendors and researchers, and ultimately create technology solutions. Once the technologies have been crafted, the standards bodies come in and begin to write recommendations that will satisfy the needs of most vendors. Meanwhile, those same vendors wait patiently on the technology sidelines.

The real world, unfortunately, is rather different than the one just described. In the real world, the technologists create the technologies that underlie such standards as SONET and SDH, and the standards bodies ultimately create standards. However, all too often the manufacturers get antsy, and in order to be the first to market with a viable product, they often jump the gun, try to interpret the early indications from the standards bodies, and design, build, and release products. In many cases, the standards process is far enough along that they interpret correctly; in other cases, they guess poorly, and the resulting products fail to meet the mark. The ultimate result of this, illustrated in Figure 3-2, is that the standards effort often gets smashed between these two great behemoths. In that case, the standards process become nothing more than a pimple on the backside of the product design process. As long as the product designers do a good job assessing the needs of the marketplace based on the work performed by the technologists

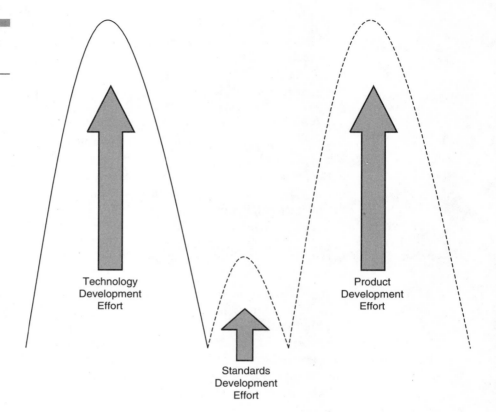

Figure 3-1
The Apocalypse of
the Two Elephants.

Technology
Development
Effort

Product
Development
Effort

Standards
Development
Effort

and the efforts of the standards bodies, the products will meet the demands of the marketplace.

ANSI's efforts to work with Bellcore and other standards and technology bodies to create an optical transmission standard were admirable. The only problem with them was that T1X1, naturally, focused its efforts on the North American marketplace (mostly the U.S.); this, after all, was the market over which it traditionally held sway. The standard that became SONET was fine for the interoperability requirements of the telephone companies in the U.S., but did little for the rest of the world. Furthermore, the deregulation of the telecommunications industry, known widely as divestiture, was a U.S. phenomenon only. The rest of the world was still tightly regulated, which meant that the proliferation of incompatible optical hardware was only a problem in the U.S. Everywhere else it was tightly controlled, as it had been in the U.S. before divestiture.

■■■ ■■■ ■■■ ■■■
Figure 3-2
The standard effort is
smashed between
the two behemoths.

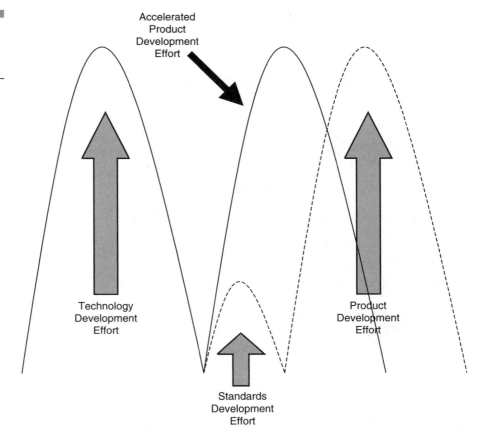

When ANSI took the SONET standard before the full ranks of the ITU, they were chagrined to learn that the ITU was not particularly motivated to give it the high-priority treatment that ANSI believed it deserved; this is because frankly, it didn't deserve it. Suddenly, SONET became something of a ship adrift.

Just before the ship hit the rocks, however, something important happened. For various reasons related to international network interoperability, standards delegates from Japan and the United Kingdom began to attend the ANSI T1X1 meetings. Suddenly, T1X1 and the European *Conference on European Post and Telegraph* (CEPT, later to be renamed ETSI) cooperated, which had the added advantage of educating the ethnocentric U.S. delegates about the needs of Europe and the rest of the world.

In early 1987, ANSI once again presented SONET to the ITU, and once again, in spite of all efforts to the contrary, it was rejected because of its North America-centric basis in technology. It suggested a bit-interleaved base transmission speed of 49.92 Mbps, which worked well for the requirements of 45-Mbps DS-3 circuits, but did nothing for the European hierarchy whose base transmission rate was 139.264 Mbps. The Europeans countered with a request that the standard be modified to accommodate a base rate of 150 Mbps so that their systems could use it. After significant give and take, the Europeans offered up a counterproposal based on what would eventually become the STS-1 standard.

The ITU then fell to the task of creating a single standard that would address the requirements of both the North American and European systems. They presented both the North American and the European proposals to the membership, and as might be expected, the North American proposal was quickly voted to second-class status. Instead of dismissing the T1X1 effort as out of hand, CEPT simply asked them to make some minor changes to it. Rather than having 49.92 Mbps as the base transmission rate, they asked that the rate be changed to 51.84 Mbps and that a byte-interleaved multiplexing technique be used instead of a bit-interleaved one. The U.S. considered the changes and agreed to them. In February of 1988, T1X1 agreed to the modifications, and SONET was on its way to becoming an internationally accepted standard. ANSI T1X1 kicked off the process with the publication of T1X1.4/87-014R4, the standard for a single-mode fiber optical interface, and T1X1.4/87-505R4, which defined transmission formats and optical transmission speeds. A new group, T1X1.5 (Optical Hierarchical Interfaces) was created to resolve any lingering comments or disputes about the newly proposed standards; a final approval was received in June of 1988. The ITU published the international version of the standards in the 1988 CCITT Blue Books as G.707, G.708, and G.709. ANSI went on to further refine the standards with the publication of T1.105-1988 and T1.106-1988, commonly referred to as the SONET Phase 1 documentation. The ITU later urged the creation of an OC-3-based international version; this enabled the rest of the world to use the standard as SDH for multiplexing the widely utilized 34-Mbps transmission signal.

I mentioned earlier that ANSI issued the Phase 1 documentation in late 1988. When the standard was first going to be published, the decision was made to publish it in phases because of the magnitude of its undertaking. Phase 1, then, addressed issues related to interoperability: frame structure, payload and overhead design, physical layer optical requirements, and payload content mappings. Phase 2 proposed OAM&P standards as well as a standard for an electrical interface to the optical SONET standard. Phase

3 added detail to the gross-level OAM&P work performed in Phase 2. It should be noted that the standards development process was originally intended to be a two-phase process, but the effort turned out to be bigger than the proverbial breadbox. Phase 2 became Phases 2 and 3, and by early 1990, the SONET standard was for all intents and purposes finally complete. Let's now turn our attention to SDH.

Why SDH?

Most SONET and SDH texts cite a common list of reasons for the proliferation of SONET and SDH networks, including a recognition of the importance of the global marketplace and a desire on the part of manufacturers, to provide devices that will operate in both SONET and SDH environments; the global expansion of ring architectures; a greater focus on network management and the value that it brings to the table; and massive, unstoppable demand for more bandwidth. These were added to those reasons: an increasing demand for high-speed routing capability to work hand-in-glove with transport; deployment of DS-1, DS-3, and E-1 interfaces directly to the enterprise customer as access solutions; growth in demand for broadband access technologies such as cable modems, the many flavors of DSL, and two-way satellite connectivity; the ongoing replacement of traditional circuit-switched network fabrics with packet-based transport and mesh architectures; a renewed focus on the SONET and SDH overhead with an eye toward using it more effectively; and the convergence of multiple applications on a single, capable, high-speed network fabric. Most visible among these is the hunger for bandwidth; according to consultancy RHK, global volume demand will grow from approximately 350,000 terabytes of transported data per month in April 2000 to more than 16 million terabytes of traffic per month in 2003. And who can argue?

Introduction: Nomenclature

Before launching into a functional description of SDH, it would be helpful to first cover the differences in naming convention between the two. This will help to dispel confusion (hopefully!).

The fundamental SONET unit of transport uses a 9-row by 90-column frame that comprises three columns of Section and Line Overhead, one column of Path Overhead, and 86 columns of payload. The payload, which is primarily user data, is carried in a payload envelope that can be formatted in various ways to make it carry a variety of payload types. For example, multiple SONET STS-1 frames can be combined to create higher-rate systems for transporting multiple STS-1 streams or a single higher-rate stream created from the combined bandwidth of the various multiplexed components. Conversely, SONET can transport sub-rate payloads, known*virtual tributaries* *virtual tributaries*, which operate at rates slower than the fundamental STS-1 SONET rate. When this is done, the payload envelope is divided into virtual tributary groups, which can in turn transport a variety of virtual tributary types.

In the SDH world, similar words apply, but they are different enough that they should be discussed. As you will see, SDH uses a fundamental transport container that is three times the size of its SONET counterpart. It is a nine-row by 270-column frame that can be configured into one of five container types, typically written C-n (where *C* means container). *n* can be 11, 12, 2, 3, or 4; they are designed to transport a variety of payload types, as shown in Figure 3-3. (The Japanese variants are also shown because they pop up periodically.) A C-11 is designed to transport a North American

Figure 3-3
Digital hierarchies.

Europe	North America	Japan	NADH	EDH	SDH Hierarchy
64 Kbps	64 Kbps	64 Kbps	DS-0	"E-0"	
	1.544 Mbps	1.544 Mbps	DS-1		C-11
2.048 Mbps				E-1	C-12
	6.312 Mbps	6.312 Mbps	DS-2		C-2
8.448 Mbps				E-2	
		32.064 Mbps			
34.368 Mbps				E-3	C-3
	44.736 Mbps		DS-3		C-3
139.264 Mbps		95.728 Mbps		E-4	C-4

NADH: North American Digital Hierarchy
EDH: European Digital Hierarchy

T-1 signal; C-12, a European E-1 signal; C-2, a DS-2; C-3, both a European 34.368-Mbps payload and North American DS-3; and C-4, a European E-4.

When an STM-1 is formatted for the transport of virtual tributaries (known as *virtual containers* in the SDH world), the payload pointers must be modified. In the case of a payload that is carrying virtual containers, the pointer is known as an *Administrative Unit type 3* (AU-3). If the payload is *not* structured to carry virtual containers, but is instead intended for the transport of higher rate payloads, then the pointer is known as an *Administrative Unit type 4* (AU-4). Generally speaking, an AU-3 is used for the transport of North American Digital Hierarchy payloads; AU-4 is used for European signal types.

The SDH Frame

To understand the SDH frame structure, it is first helpful to understand the relationship between SDH and SONET. Functionally, they are identical. In both cases, the intent of the technology is to provide a globally standardized transmission system for high-speed data. SONET is indeed optimized for the T-1-heavy North American market, whereas SDH is more applicable to Europe; beyond that, however, the overhead and design considerations of the two are virtually identical. Nomenclature aside, the reader will begin to get a sense of déjà vu in this chapter. Some differences will certainly present themselves, and they should not be ignored; all in all, however, the similarities between the two far outweigh the differences.

One of the key items that should be mentioned right up front stems from the cross-border relationship that exists between the two standards. Because SDH is perceived by the majority of players to be the true international standard for optical transport (as opposed to SONET), a network tie always goes to SDH. In other words, on an international link connecting a SONET environment on one end to an SDH environment on the other, SDH wins: all encoding for transport must follow the more widely accepted international SDH rules of engagement.

Perhaps the greatest difference between the two lies in the physical nature of the frame. A SONET STS-1 frame comprises 810 total bytes, for an overall aggregate bit rate of 51.84 Mbps—perfectly adequate for the North American 44.736-Mbps DS-3. An SDH STM-1 frame, however, designed to transport a 139.264-Mbps European E-4 or CEPT-4 signal, must be larger if it is to accommodate that much bandwidth—it clearly won't fit in the limited space available in an STS-1. An STM-1, then, operates at a fundamental rate of

155.52 Mbps, enough for the bandwidth requirements of the E-4. This should be where the déjà vu starts to kick in. Perceptive readers will remember the 155.52 Mbps number from our discussions of the SONET STS-3, which offers *exactly* the same bandwidth. Figure 3-4 shows an STM-1 frame. It is a byte-interleaved, nine-row by 270-column frame, with the first nine columns devoted to overhead and the remaining 261 devoted to payload transport.

A comparison of the bandwidth between SONET and SDH systems is also interesting. The fundamental SDH signal is exactly *three times* the bandwidth of the fundamental SONET signal, and this relationship continues all the way up the hierarchy, as shown in Figure 3-5. STM-256 is shown parenthetically because it is largely in trial only at the time of this writing, although its arrival on the market is imminent.

STM Frame Overhead

The overhead in an STM frame is very similar to that of an STS-1 frame, although the nomenclature varies somewhat. Instead of Section, Line, and

Figure 3-4
STM-1 frame.

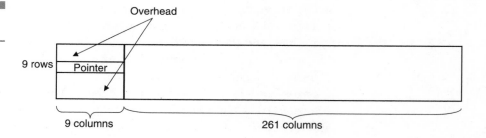

Figure 3-5
Comparison of SDH and SONET.

SDH Level	SONET Level	Bandwidth
STM-1	OC-3	155.52 Mbps
STM-16	OC-48	2.488 Gbps
STM-64	OC-192	9.954 Gbps
(STM-256)	OC-768	39.81 Gbps

Path Overhead to designate the different regions of the network that the overhead components address, SDH uses *Regenerator Section*, *Multiplex Section*, and *Path* Overhead, as shown in Figure 3-6. The *Regenerator Section Overhead* (RSOH) occupies the first three rows of nine bytes, and the *Multiplex Section Overhead* occupies (MSOH) the final five. Row four is reserved for the pointer. As in SONET, the Path Overhead floats gently on the payload tides, rising and falling in response to phase shifts. Functionally, these overhead components are identical to their SONET counterparts; the table in Figure 3-7 illustrates the two.

Overhead Details

Because an STM-1 is three times as large as an STS-1, it has three times the overhead capacity—nine columns instead of three (plus Path Overhead). This overhead is shown in Figure 3-8 and described in the following section; an STS-3c frame is shown in Figure 3-9 for comparative purposes.

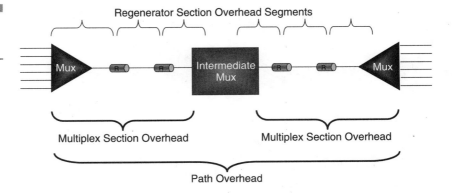

Figure 3-6
Regions of the SDH network.

Figure 3-7
SDH vs. SONET nomenclature.

SDH Nomenclature	SONET Nomenclature
Regenerator Section Overhead	Section Overhead
Multiplex Section Overhead	Line Overhead
Path Overhead	Path Overhead

Figure 3-8
STM overhead.

A1	A1	A1	A2	A2	A2	J0	RNU	RNU	J1
B1	MDU	MDU	E1	MDU	R	F1	RNU	RNU	B3
D1	MDU	MDU	D2	MDU	R	D3	R	R	C2
				AU-n Pointer					G1
B2	B2	B2	K1	R	R	K2	R	R	F2
D4	R	R	D5	R	R	D6	R	R	H4
D7	R	R	D8	R	R	D9	R	R	F3
D10	R	R	D11	R	R	D12	R	R	K3
S1	Z1	Z1	Z2	Z2	M1	E2	RNU	RNU	N1

RNU: Reserved for National Use
MDU: Media-Dependent Use
R: Reserved

Figure 3-9
STS-3c frame.

A1	A1	A1	A2	A2	A2	J0	R	R	J1
B1	R	R	E1	R	R	F1	R	R	B3
D1	R	R	D2	R	R	D3	R	R	C2
H1	H1	H1	H2	H2	H2	H3	H3	H3	G1
B2	B2	B2	K1	R	R	K2	R	R	F2
D4	R	R	D5	R	R	D6	R	R	H4
D7	R	R	D8	R	R	D9	R	R	F3
D10	R	R	D11	R	R	D12	R	R	K3
S1	Z1	Z1	Z2	Z2	M1	E2	R	R	N1

(It might be helpful to review the layout of a concatenated SONET frame before proceeding.) Notice the subtle but important differences in the overhead of the two. The first row of the RSOH is its SONET counterpart, with the exception of the last two bytes, which are labeled as being reserved for national use and are specific to the PTT administration that implements the network. In SONET they are not yet assigned. The second row is different from SONET in that it has three bytes reserved for media-dependent implementation (differences in the actual transmission medium, whether copper, coaxial, or fiber) and the final two reserved for national use. As before, they are not yet definitively assigned in the SONET realm.

The final row of the RSOH also sports two bytes reserved for media-dependent information, as they are reserved in SONET. All other RSOH functions are identical between the two.

The MSOH in the SDH frame is almost exactly the same as that of the SONET Line Overhead, with one exception: row nine of the SDH frame has two bytes reserved for national administration use. They are also reserved in the SONET world.

The pointer in an SDH frame is conceptually identical to that of a SONET pointer, although it has some minor differences in nomenclature. In SDH, the pointer is referred to as an *Administrative Unit* (AU) pointer, referring to the standard naming convention described earlier.

Regenerator Section Overhead

The RSOH occupies the first three rows of the first nine columns of an STM frame. The bytes contained in this overhead and their functions are described in the following section and shown in Figure 3-10.

A1, A2: The A1 and A2 bytes provide a unique framing pattern at the beginning of the frame that is used to identify the beginning of the frame to receiving equipment for synchronization purposes. The pattern is the hexadecimal number 0xF628, which in binary is 1111 0110 0010 1000. When transmitting STM-n frames, the A1 and A2 bytes must appear in the overhead of each STM-1 in the STM-n frame.

J0: A unique number called the STM-1 ID identifies every STM-1. It is carried in the J0 byte and is used to identify the originating office of a frame

Figure 3-10
Regenerator Section
Overhead.

A1	A1	A1	A2	A2	A2	J0	RNU	RNU
B1	MDU	MDU	E1	MDU	R	F1	RNU	RNU
D1	MDU	MDU	D2	MDU	R	D3	R	R

as it makes its way across the network. Because SDH transmission can involve multiple levels of interleaved multiplexing, the J0 byte provides a way to uniquely identify each frame of data.

B1: The B1 byte is known as the *Bit-Interleaved Parity* byte (BIP-8). It provides an 8-bit parity check that is used for error detection. In practice, the parity check is performed across all bytes of the prior frame and placed in the B1 byte of the current frame prior to scrambling.

E1: The Orderwire byte is a 64-Kbps voice channel that can be used by technicians to communicate while troubleshooting between repeater spans. When building STM-n frames, the E1 byte is only defined in the first STM of the STM-n for obvious reasons.

F1: The F1 byte, called the user byte, is user configurable and can be employed for a variety of purposes. It is reserved for the use of the service provider as a transport byte for network management application information, although many service providers today use it to transport maintenance communications and configuration data. Because the byte's specific use is not specified in a standard, it is extremely flexible, but lends very little to the overall goal of interoperability.

D1, D2, D3: The *Data Communications Channel* (DCC) bytes provide a 192-Kbps communications channel between section-terminating devices and are used to transport *operations, administration, and maintenance* (OAM) information such as control signals, monitoring, alarm information, and so on. Although the DCC bytes are fully in alignment with the nascent *Telecommunications Management Network* (TMN) standards that continue to evolve, it will probably be some time before their use is clearly defined among operators to ensure interoperability.

Multiplex Section Overhead

The MSOH comprises the final four rows of the first nine columns of the STM-1, as shown in Figure 3-11. These overhead bytes are described in the following section.

B2: B2 is another bit-interleaved parity byte and is used to carry error-checking information. It calculates its value based on the MSOH (it does not include the RSOH) and payload of the previous frame before it is scrambled for transmission. The information is then placed in the current frame before it is scrambled. The BIP-8 is required for all STMs in an STM-n.

K1/K2: Like SONET, one of the principal advantages of SDH is its capability to detect a failure in the network and switch traffic to a backup trans-

Figure 3-11
Multiplex Section
Overhead.

B2	B2	B2	K1	R	R	K2	R	R
D4	R	R	D5	R	R	D6	R	R
D7	R	R	D8	R	R	D9	R	R
D10	R	R	D11	R	R	D12	R	R
S1	Z1	Z1	Z2	Z2	M1	E2	RNU	RNU

mission span automatically if one is available, ensuring protection of cus-tomer data. This technique is called *automatic protection switching*. If the network is deployed across a ring architecture, the K1 and K2 bytes are used to switch from one ring to another.

D4-D12: These nine bytes comprise a 576-Kbps message channel that SDH network elements can use to transport network management (OAM) messages. The information carried in these bytes is similar to that carried in the Regenerator Section Layer's Data Com channels, but in this case is specific to the Multiplex Section Layer. These bytes are only defined for the first STM in an STM-n. Like the Regenerator Section Overhead's DCC bytes, they are designed to use TMN protocols.

S1/Z1: This byte is still in development in terms of its specific responsi-bilities. Of its eight bits, the last four (bits five through eight) have been reserved for the transport of timing information that permits a network ele-ment to choose a specific clock source based on its own selection parameters as a way to prevent timing problems within the network. This byte is defined in the first STM of an STM-n only; all remaining S1 bytes are des-ignated as Z (growth) bytes.

M0/M1/Z2: This particular byte is something of a chameleon in that it can take on a number of roles depending on the nature of the payload with which it is associated. If the payload being transported is a single STM-1, then the byte becomes the M0 byte and is used for error control and is called the *Remote Error Indication-Line Level* (REI-L). In earlier versions of SDH, one of the signals that was closely monitored was called a *Far-End Block Error* (FEBE), which indicated the presence of bit-level errors. It has been replaced by the REI-L, and its job is to transport the BIP-8 informa-tion back to the presumed source of the problem. The last four bits of the M0 byte are used for this function; the first four are undefined. If the byte is not used for M0 or M1 functionality, it is designated as a growth byte (Z2).

E2: Orderwire: The E2 byte is functionally identical to the E1 byte discussed in the RSOH section. It provides a standard 64-Kbps voice channel for maintenance communications between technicians and is only defined in the first STM of an STM-n.

AU-n pointer: As described earlier, the payload pointer in the SDH network world is called an AU and comes in two flavors—either an AU-3 or an AU-4. If the payload being transported is a VC-3, then the pointer is an AU-3; if the payload is a VC-4, then the pointer is an AU-4. Remember that the virtual containers are analogous to the payload envelope.

When SONET and SDH systems must interoperate, some of the differences in the pointers must be taken into account. The SDH pointer, for example, defines bits five and six as *size bits*, which is different from their role in SONET pointers. They must therefore not be interpreted by SONET receiving devices. In the real world, most SONET devices simply ignore them, but awareness of the differences is still important.

Path Overhead

As shown in Figure 3-12, the SDH Path Overhead comprises nine bytes, just as it does in SONET networks. These bytes include J1, B3, C2, G1, F2, H4, F3, K3, and N1. Each is described in the following section.

J1: The path trace byte is used by a receiving device to ensure that it is connected to the proper transmitting device. The field is used to generate a 64-byte repeating string that identifies the source of the received signal. The field can be programmed by the user any way he or she likes and often

Figure 3-12
Path Overhead.

J1
B3
C2
G1
F2
H4
F3
K3
N1

contains an IP address, a *Common Language Location Code* (CLLC, or *click code*), a *Common Language Location Indicator* (CLLI, or *silly code*), or a telephone number that identifies the transmitting device. In SDH systems, the J1 byte often contains an E-164 because that is the only address format that is allowed at international handoff points. If for some reason neither party uses the field, it is filled with null characters.

B3: The Path BIP-8 is used to error-check the content of the previous payload before being scrambled.

C2: The signal label byte is used to tell a receiving device what is actually contained in the payload that is arriving in terms of the actual construction of the payload. This permits the simultaneous transport of multiple traffic types. This byte is one of the most important in the SDH domain, particularly for SDH-SONET system handoffs. Figure 3-13 illustrates the differences that exist in payload mappings between SONET and SDH systems.

Figure 3-13
Payload mappings in SDH and SONET.

Hex Code	SDH Payload Mapping (C2)	SONET Payload Mapping (C2)
00	Unequipped or supervisory	Unequipped; not available for live traffic
01	Equipped – non-specific	Equipped – non-specific
02	TUG structure (VCs inside)	VT structure (VTs inside)
03	TUs in locked mode	VTs in locked mode (no longer supported)
04	Asynchronous E-3 or DS-3	DS-3 mapping inside C-3
12	Asynchronous E-4	Asynchronous DS-4NA mapping: C-4
13	ATM cell mapping	ATM cell mapping
14	DQDB cell mapping	DQDB cell mapping
15	Asynchronous FDDI mapping	Asynchronous FDDI mapping
16	HDLC over SDH (for IP transport)	HDLC over SONET (for IP transport)
CF	Experimental or trial for IP in PPP	Experimental or trial for IP in PPP
FE	Q.181 test signal	G.707 test signal mapping
FF	Virtual container alarm	Not defined

G1: As its name implies, the path status byte is used to communicate the overall transmission status of the duplex circuit to the originating device. In practice, it transports two indicators: the *Path Remote Error Indicator* (REI-P) in bits one through four, and the *Path Remote Defect Indicator* (RDI-P) in bits five through seven, leaving a single unused bit.

F2/3: The network service provider can use these bytes to transport communications information such as network management data.

H4: This multiframe indicator byte is used to indicate the relative position of payload within a container and is often specific to the payload being transported. It is used when payloads are configured as virtual containers and four of them are grouped into a superframe or multiframe.

K3: Formerly known as the Z4 growth byte, the F3 byte has been reassigned for future use.

N1: Formerly known as the Z5 growth byte, N1 is used for the transport of network maintenance and management information.

This concludes our discussion of the SDH overhead bytes. We now turn our attention to the SDH tributary units designed to transport sub-rate payloads.

SDH Tributary Units

Similar to what we saw with SONET virtual tributaries, SDH tributary units are designed to carry payloads that operate at speeds lower than the SDH base rate. In fact, the SDH and SONET tributary units are perfectly aligned with each other, as shown in Figure 3-14.

Like the virtual tributaries of SONET, SDH tributary units are ganged together to form a multiframe for transmission. Each begins with an overhead byte (V1, V2, V3, or V4), and naturally the multiframe begins with the

Figure 3-14
SDH tributary units.

TU Format	#Columns	Bytes/Frame	Bandwidth	Payload
TU-11	3	27	1.728 Mbps	DS-1
TU-12	4	36	2.304 Mbps	E-1
TU-2	12	108	6.912 Mbps	DS-2

V1 byte. The Multiframe Indicator Octet (H4) indicates the phase of the multiframe signal, which helps receiving devices locate the payload upon receipt.

A single STM-1 can transport 84 TU-11s, similar to what we saw with SONET. Byte-interleaved transport is used in these systems, so alignment and the role of the AU-4 pointer are important. As in SONET, the locations of the tributary data are fixed and well known, and therefore relatively easy to find.

Figure 3-15, adapted from Goralski, shows the compatible payload mappings between SONET and SDH. The table serves as a useful tool for comparing the two environments.

We have now covered both SONET and SDH in some detail. In the next few sections we will discuss corollary technologies, including optical transport and ATM.

Figure 3-15

Payload mapping comparison in SDH and SONET.

Payload	Bandwidth	In STS-1	In STS-Nc	In AU-3	In AU-4
DS-1	1.544 Mbps	VT1.5	NA	VC-11/12	VC-11/12
E-1	2.048 Mbps	VT2	NA	VC-12	VC-12
DS-1C	3.152 Mbps	VT3	NA	NA	NA
DS-2	6.312 Mbps	VT6	NA	VC-2	VC-2
E-3	34.368 Mbps	NA	NA	VC-3	VC-3
DS-3	44.736 Mbps	STS-1	NA	VC-3	VC-3
FDDI	125 Mbps	NA	STS-3c	NA	C-4
DS-4NA	139.264 Mbps	NA	STS-3c	NA	VC-4
DQDB	149.760 Mbps	NA	STS-3c	NA	C-4
ATM Cells	149.760 Mbps	STS-1	STS-3c	VC-3	VC-4

NA: Not applicable.

Overview of Optical Technology

In the final analysis, SONET is a physical layer standard for the transmission of high-speed, multiplexed data across an optical network. This section discusses the fundamentals of optical network technologies, including the basics of fiber optics, the fundamentals of optical networking, common transmission impairments, *Dense Wavelength Division Multiplexing* (DWDM), and emerging optical switching and routing technologies.

Early Technology Breakthroughs

In 1878, two years after perfecting his speaking telegraph (which became the telephone), Alexander Graham Bell created a device that transmitted the human voice through the air for distances up to 200 meters. The device, which he called the Photophone, used carefully angled mirrors to reflect sunlight onto a diaphragm that was attached to a mouthpiece, as shown in Figure 4-1. At the receiving end (Figure 4-2), the light was concentrated by a parabolic mirror onto a selenium resistor, which was connected to a battery and speaker. The diaphragm vibrated when struck by the human voice, causing the intensity of the light striking the resistor to vary. The selenium resistor, in turn, caused the current flow to vary in concert with the varying sunlight, causing the received sound to come out of the speaker with remarkable fidelity. This represented the birth of optical transmission.

Figure 4-1
Photophone
transmitter.

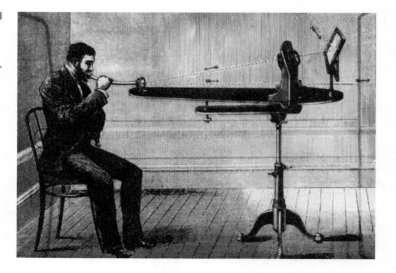

Figure 4-2
Photophone receiver.

Earlier demonstrations of optical transmissions were done before this. Swiss physicist Daniel Colladon (in 1841), and later in 1870, physicist John Tyndall (well known for his work on the properties of gases) demonstrated that a beam of light would follow (for the most part) a stream of water issuing from a container, showing that the air-water interface would reflect most of the light back into the stream. Ten years after Tyndall, William Wheeler of Concord, Massachusetts (who later became a well-known hydraulics engineer, of all things), created a practical application for Tyndall's demonstration when he used highly reflective metal pipes to carry the brilliant light from a carbon arc lamp to various rooms in a house. His technique never proved to be commercially practical, but it was the first attempt to pump light for practical reasons, and was the first demonstration of an actual lightguide—a concept that would later be perfected with the development of fiber optics. Figure 4-3 shows a sketch of Wheeler's invention.

It is clear that our fascination with pumped light stems from a point deep in the annals of history, but it has only been relatively recently that optical science has been perfected to the point that optical transmission and its corollary optical switching technology have become not only marketable, but are in fact redefining the nature of data networking.

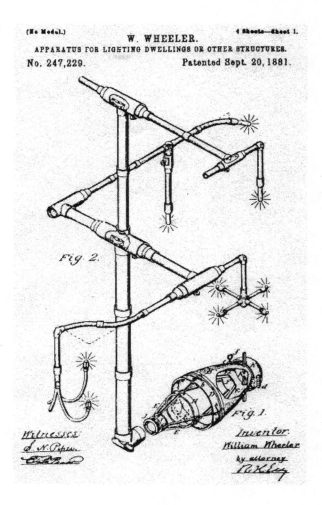

Total Internal Reflection

In each of their experiments, Colladon, Wheeler, and Tyndall relied on a phenomenon called *total internal reflection*, which is fundamental to understanding how optical transmission works. Therefore, we must delve into optical physics for a moment. This will only be slightly uncomfortable, so go with me on this.

Everyone at one time or another has seen the image of a stick appearing to bend when it is inserted in water, or the frustration of a hunter trying to spear a fish, only to discover that the fish isn't where it appears to be. This phenomenon, called *refraction*, occurs because of a difference in the *refractive index* between the air and the water. The refractive index is a measure of the ratio between the speed of light in a vacuum (actually, today it is measured through the air) and the speed of light in the other medium. Light travels slower in physical media than it does when transmitted through the air, so given that the refractive index [n] is measured as

$$\frac{\text{Speed of light [c] in a vacuum}}{\text{Speed of light [c] in another medium,}}$$

the refractive index for any other medium will always be *greater* than 1.

So why do we care about this? Because the light actually bends when it passes through the interface between media with different refractive indices. So for example, when a light source shines a beam of light into a glass fiber, the light bends as it passes from the air into the glass. The degree that it bends is a function of two things: the difference in refractive index between the two media, and the angle at which the light strikes the glass, known as the *angle of incidence*. This angle is measured from the centerline of the medium, a line that runs perpendicular to the entry surface. For fiber optic transmission systems, this becomes rather crucial. Figure 4-4 illustrates this concept.

The relationship between the angle of incidence and the angle of refraction is called *Snell's Law*. It becomes important in fiber systems because of the criticality of having the light enter the fiber from the source at as narrow an angle of incidence as possible. If the angle of incidence is too high, as shown in Figure 4-5, the light can actually escape from the glass, resulting in severe signal loss. According to Snell's Law, if the angle of incidence is too high, then refraction will not take place. Put another way, if the light strikes the interface between air and glass (passing into a material with a higher refractive index) at a steep enough angle, the light will not escape,

Figure 4-4
Angle of incidence and angle of refraction.

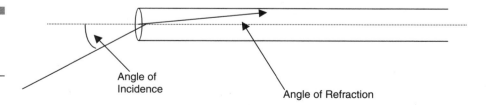

Angle of Incidence

Angle of Refraction

but will be reflected back into the glass instead. This process, shown in Figure 4-6, is called *total internal reflection*. It is the basis for transmission through optical fiber. The more light that can be kept inside the fiber, the better the integrity (power) of the transmitted signal.

Consequently, the angle of the incident light impinging upon the face of the fiber, often called the *acceptance angle* or *numerical aperture*, is rather important if the signal is to be transmitted over any significant distance. Therefore, the laser assemblies that serve as the source of the transmitted light must be carefully crafted to ensure that the face of the laser that will generate the signal to be transmitted is aligned as closely as possible with the face of the fiber—particularly the core of the fiber where the light actually travels. Figure 4-7 illustrates this process. When we consider that modern single-mode fiber has a core diameter of approximately 8 microns, then we realize that the laser itself must be approximately that diameter if it is to direct the bulk of its light output down the core of the fiber. This takes on meaning when we note that a human hair has a diameter of approximately 50 microns!

Even in the best systems, about 4 percent of the signal is lost at these air/glass interfaces between the laser and the face of the fiber core. This loss is known in the industry as *Fresnel Loss*.

Figure 4-5
Angle of incidence
too high.

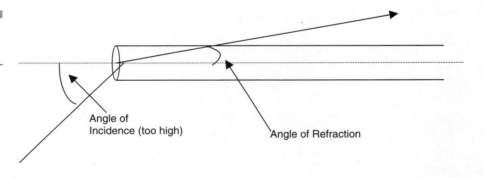

Figure 4-6
Total internal
reflection.

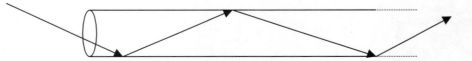

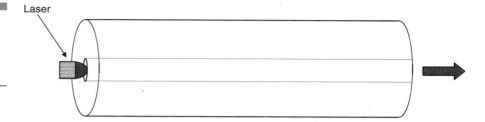

Figure 4-7
Critical relationship between laser light source and fiber core.

Laser

Because of its design, optical fiber serves as a nearly perfect light guide. It consists of two layers: an inner core through which the bulk of the light travels, and an outer cladding that serves to keep the light in the core, as shown in Figure 4-8.

This is accomplished by taking advantage of Snell's Law. In typical optical fiber, the refractive index of the core is slightly—but only slightly—higher than the refractive index of the cladding. Thus, the two refractive indices, working closely with the angle of incidence of the transmitted light, ensure that minimal light escapes from the core. If it were not for the cladding, much of the light would escape from the core and be lost over distance.

All right, enough physics for a moment—back to the industry.

Developments in Optical Transmission

Transmission of light across a bounded medium saw its first practical application with the creation of the fiberscope, simultaneously invented in the 1950s by Narinder Kapany at the Imperial College of Science and Technology in London, and Brian O'Brian of the American Optical Company. The device, designed to conduct an optical image from a source to a destination, relied on glass fiber and was used for industrial inspections and medical applications. Luckily, the distances traversed were quite short because the unclad fibers of the fiberscope suffered tremendous loss, even over such short distances.

The next major advancement in optical transmission came with the creation of high-quality, low-cost light sources. These came initially in two forms: as *light-emitting diodes* (LEDs) and as *laser diodes* (LDs). Gordon Gould, a graduate student at Columbia University, did the initial work on

Figure 4-8
Optical fiber
anatomy.

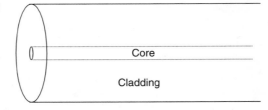

coherent laser light in the 1950s; researchers at Bell Laboratories in Murray Hill, New Jersey took it to the next level by attempting to craft practical applications for the technology. In 1962, the first semiconductor lasers were created; these became the focal point for transmission over fiber optics.

Of course, optical transmission in the early days was limited in its capabilities. Even though modulated light has an information carrying capacity that is orders of magnitude greater than radio, its earliest incarnations suffered tremendous signal loss over distance because of impurities in the glass and the limitations of the optoelectronics that drove the light signal.

Consider the following analogy. If you look through a 2-foot square pane of window glass, it appears absolutely clear—if the glass is clean, it is virtually invisible. However, if you turn the pane on edge and look through it from edge to edge, the glass appears to be dark green. Very little light passes from one edge to the other. In this example, you are looking through two feet of glass. Imagine trying to pass a high-bandwidth optical signal through 40 or more kilometers of that glass!

In 1966, Charles Kao and Charles Hockham at the U.K.'s Standard Telecommunication Laboratory (now part of Nortel Networks) published their seminal work, demonstrating that optical fiber could be used to carry information, provided its end-to-end signal loss could be kept below 20 dB per kilometer. Keeping in mind that the decibel scale is logarithmic, 20 dB of loss means that 99 percent of the light would be lost over each kilometer of distance. Only 1 percent would actually reach the receiver—and that's only a 1-km run. Imagine the loss over today's fiber cables that are hundreds of kilometers long, if 20 dB was the modern performance criterion!

Kao and Hockham proved that metallic impurities in the glass such as chromium, vanadium, iron, and copper were the primary cause for such high levels of loss. In response, glass manufacturers rose to the challenge and began to research the creation of ultrapure products.

In 1970, Peter Schultz, Robert Maurer, and Donald Keck of Corning Glass Works (now Corning Corporation) announced the development of a glass fiber that offered better attenuation than the recognized 20-dB threshold. Today, fiber manufacturers offer fiber so incredibly pure that 10 percent of the light arrives at a receiver placed 50 kilometers away. Put another way, a fiber with 0.2 dB of measured loss delivers more than 60 percent of the transmitted light over a distance of 10 kilometers. Remember the windowpane example? Imagine glass so pure that you could see clearly through a window 10 kilometers thick.

Fundamentals of Optical Networking

At their most basic level, optical networks require three fundamental components (shown in Figure 4-9): a source of light, a medium over which to transport it, and a receiver for the light. Additionally, regenerators, optical amplifiers, and other pieces of equipment in the circuit may be included. We will examine each of these generic components.

Optical Sources

Today, the most common sources of light for optical systems are either light-emitting diodes or laser diodes. Both are commonly used, although laser diodes have become more common for high-speed data applications because of their coherent signal. Although lasers have gone through several iterations over the years including ruby rod and helium-neon, semiconductor lasers became the norm shortly after their introduction in the early 1960s because of their low cost and high stability.

Figure 4-9

Components of a typical optical network span.

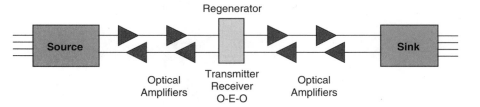

Light-Emitting Diodes (LEDs)

Light-emitting diodes come in two varieties: *surface-emitting LEDs* and *edge-emitting LEDs*. Surface-emitting LEDs, illustrated schematically in Figure 4-10, give off light at a wide angle, and therefore do not lend themselves to the more coherent requirements of optical data systems because of the difficulty involved in focusing their emitted light into the core of the receiving fiber. Instead, they are often used as indicators and signaling devices. They are, however, quite inexpensive and are therefore commonly found in more forgiving applications.

An alternative to the surface-emitting LED is the edge-emitting device, shown in Figure 4-11. Edge emitters produce light at significantly narrower angles and have a smaller emitting area, which means that more of their emitted light can be focused into the core. They are typically faster devices than surface emitters, but do have a downside: they are temperature-sensitive, and must therefore be installed in environmentally controlled devices to ensure the stability of the transmitted signal.

Figure 4-10
Surface emitting laser.

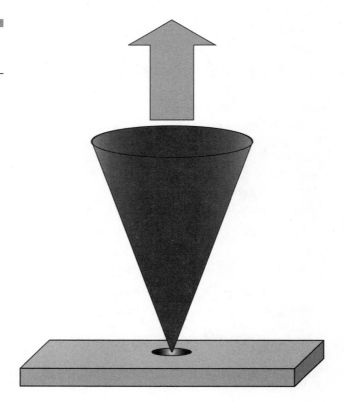

Figure 4-11
Edge emitting laser.

Laser Diodes

Laser diodes represent the alternative to LEDs. A laser diode has a very small emitting surface, usually no larger than a few microns in diameter, which means that a great deal of the emitted light can be directed into the fiber. Because they represent a coherent source, the emission angle of a laser diode is extremely narrow. It is the fastest of the three devices.

Many different types of laser diodes are available. The most common are the *electro-absorptive modulated laser* (EML), which combines a continuous wave laser with a modulating shutter device; the *distributed feedback laser*, which has an integrated grating assembly to maintain a constant output frequency; a *vertical cavity surface-emitting laser* (VCSEL, pronounced "vick-sel"), which produces light from a round spot, resulting in a beam of light that is less prone to spread than a typical surface-emitting laser's output. VCSELs are low-power, low-cost, multifrequency devices. Finally, *Fabry-Perot lasers* are older devices that suffer a number of problems and are less commonly used. They tend to emit light at multiple, closely spaced wavelengths and are commonly called multimode lasers.

Figure 4-12 schematically shows the emission characteristics of all three devices. The surface-emitting LED has the widest emission pattern, followed by the edge emitter; the laser diode represents the most coherent and therefore effective light generator. In fact, the graph of the output signal of an LED versus that of a laser is rather dramatic, as shown in Figure 4-13 (vertical axis not to scale).

Optical Fiber

When Peter Schultz, Donald Keck, and Robert Maurer began their work at Corning to create a low-loss optical fiber, they did so using a newly crafted process called *inside vapor deposition* (IVD). Whereas most glass is manufactured by melting and reshaping silica, IVD deposits various combinations of carefully selected compounds on the inside surface of a silica tube. The tube becomes the cladding of the fiber; the vapor-deposited compounds

Figure 4-12
Emission patterns
of most common
light sources.

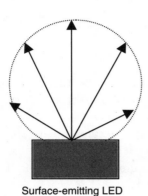

Surface-emitting LED

Edge-emitting LED

Laser Diode

Figure 4-13
Representative pulse
widths of laser, LED
(not to scale).

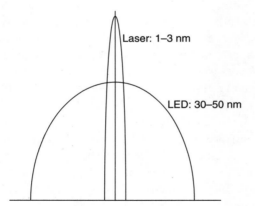

Laser: 1–3 nm

LED: 30–50 nm

become the core. The compounds are typically silicon chloride ($SiCl_4$) and oxygen (O_2), which are reacted under heat to form a soft, sooty deposit of silicon dioxide (SiO_2), as illustrated in Figure 4-14. In some cases, impurities such as germanium are added at this time to cause various effects in the finished product. In practice, the $SiCl_4$ and O_2 are pumped into the fused silica tube as gases; the tube is heated in a high-temperature lathe, causing

■■ ■■ ■■ ■■

Figure 4-14
Manufacturing fiber
using internal vapor
deposition (IVD)
process.

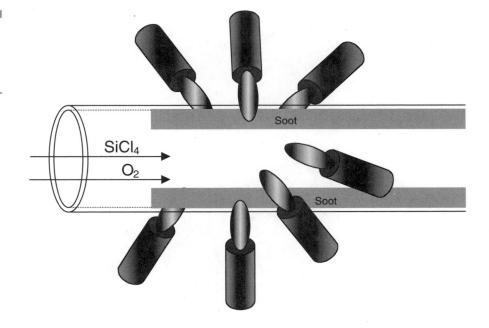

the sooty deposit to collect on the inside surface of the tube (Figure 4-14). The continued heating of the tube causes the soot to fuse into a glass-like substance.

This process can be repeated as many times as required to create a graded refractive index. Ultimately, once the deposits are complete, the entire assembly is heated fiercely, which causes the tube to collapse, creating what is known in the optical fiber industry as a *preform*. Figure 4-15 shows an example of a preform.

An alternative manufacturing process is called *outside vapor deposition* (OVD). In the OVD process, the soot is deposited on the surface of a rotating ceramic cylinder in two layers (Figure 4-16). The first layer is the soot that will become the core; the second layer becomes the cladding. Ultimately, the rod and soot are sintered to create a preform. The ceramic is then removed, leaving behind the fused silica that will become the fiber.

A number of other techniques are available for creating the preforms that are used to create fiber, but these are the principal techniques in use today. The next step is to convert the preform into optical fiber.

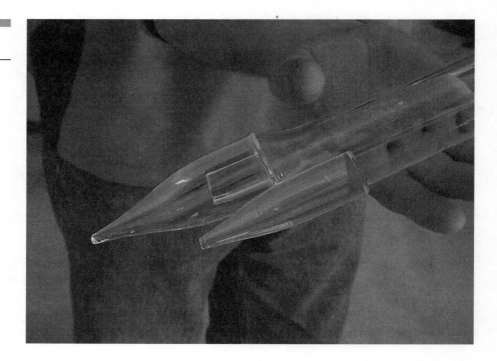

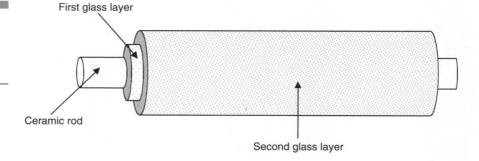

First glass layer

Ceramic rod

Second glass layer

Drawing the Fiber

To make fiber from a preform, the preform is mounted in a furnace, shown in Figure 4-17, at the top of a tall building called a *drawing tower*. The bottom of the preform is heated until it has the consistency of taffy, at which time the soft glass is drawn down to form a thin fiber. When it strikes the cooler air outside the furnace, the fiber solidifies. Needless to say, the process is carefully managed to ensure that the thickness of the fiber is precise; microscopes such as the one in Figure 4-18 are used to verify the geometry of the fiber.

Other stages in the manufacturing process include a monitoring process to check the integrity of the product, a coating process that applies a protective layer, and a take-up stage where the fiber is wound onto reels for later assembly into cables of various types.

Optical Fiber

Dozens of different types of fiber are available. Some of them are holdovers from previous generations of optical technology that are still in use and represented the best efforts of technology available at the time; others represent improvements on the general theme or specialized solutions to specific optical transmission challenges.

Figure 4-17
Drawing furnaces.

Figure 4-18
Fiber manufacturing
quality control
process.

Figure 4-18
Fiber manufacturing
quality control
process.

Generally speaking, two major types of fiber are used: *multimode,* which is the earliest form of optical fiber and is characterized by a large diameter central core, short-distance capability, and low bandwidth; and *single mode,* which has a narrow core and is capable of greater distance and higher bandwidth. Varieties of each of these will be discussed in detail later in the book.

To understand the reason for and philosophy behind the various forms of fiber, it is first necessary to understand the issues that confront transmission engineers who design optical networks.

Optical fiber has a number of advantages over copper: it is lightweight, has enormous bandwidth potential, has significantly higher tensile strength, can support many simultaneous channels, and is immune to electromagnetic interference. It does, however, suffer from several disruptive problems that cannot be discounted. The first of these is *loss* or *attenuation,* the inevitable weakening of the transmitted signal over distance that has a direct analog in the copper world. Attenuation is typically the result of two sub-properties. The first is *scattering* and *absorption,* both of which have cumulative effects, and the second is *dispersion*, which is the spreading of the transmitted signal and is analogous to noise.

Scattering

Scattering occurs because of impurities or irregularities in the physical makeup of the fiber itself. The best-known form of scattering is called *Rayleigh Scattering*; it is caused by metal ions in the silica matrix and results in light rays being scattered in various directions, as illustrated in Figure 4-19.

Rayleigh Scattering occurs most commonly around wavelengths of 1,000 nm and is responsible for as much as 90 percent of the total attenuation that occurs in modern optical systems. It occurs when the wavelengths of the light being transmitted are roughly the same size as the physical molecular structures within the silica matrix; thus, short wavelengths are affected by Rayleigh Scattering effects far more than long wavelengths. In fact, it is because of Rayleigh Scattering that the sky appears to be blue: the shorter (blue) wavelengths of light are scattered more than the longer wavelengths of light.

Absorption

Absorption results from three factors: hydroxyl (OH^-; water) ions in the silica; impurities in the silica; and incompletely diminished residue from the

Figure 4-19

Scattering effect in optical fiber.

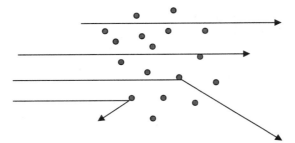

manufacturing process. These impurities tend to absorb the energy of the transmitted signal and convert it to heat, resulting in an overall weakening of the optical signal. Hydroxyl absorption occurs at 1.25 and 1.39 μm; at 1.7 μm, the silica itself starts to absorb energy because of the natural resonance of silicon dioxide.

It is interesting to take a side road for a moment and examine hydroxyl absorption because it is the basis for the proper functionality of a commonly used household device: the microwave oven. If you place a dry glass measuring cup in a microwave oven, you can cook it in the oven for an hour and it will barely be warm to the touch. Fill it with water, however, or wet it lightly, and it will rapidly reach the boiling point of water. This happens because microwave ovens generate microwave energy at a wavelength that is effectively absorbed by the hydroxyl ions in water. The hydroxyl ions convert the energy to heat, the water boils, the meat cooks, and the potatoes bake. This is also the reason that microwave transmission systems tend to experience serious signal degradation during heavy rain and fog: the water in the air between the transmitter and the receiver absorb the microwave energy, resulting in a weakened signal—sometimes dramatically so.

Dispersion

As mentioned earlier, dispersion is the optical term for the spreading of the transmitted light pulse as it transits the fiber. It is a bandwidth-limiting phenomenon and comes in two forms: *multimode dispersion* and *chromatic dispersion*. Chromatic dispersion is further subdivided into *material dispersion* and *waveguide dispersion*.

Multimode Dispersion To understand multimode dispersion, it is first important to understand the concept of a *mode*. Figure 4-20 shows a fiber

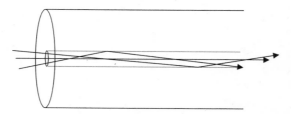

Figure 4-20
Wide core
optical fiber.

with a relatively wide core. Because of the width of the core, it enables light rays arriving from the source at a variety of angles (three in this case) to enter the fiber and be transmitted to the receiver. Because of the different paths that each ray, or mode, will take, they will arrive at the receiver at different times, resulting in a dispersed signal.

Now consider the system shown in Figure 4-21. The core is much narrower and only allows a single ray, or mode, to be sent down the fiber. This results in less end-to-end energy loss and avoids the dispersion problem that occurs in multimode installations.

Chromatic Dispersion   The speed at which an optical signal travels down a fiber is absolutely dependent upon its wavelength. If the signal comprises multiple wavelengths, then the different wavelengths will travel at different speeds, resulting in an overall spreading or smearing of the signal. As discussed earlier, chromatic dispersion comprises two subcategories: material dispersion and waveguide dispersion.

Material Dispersion   Simply put, material dispersion occurs because different wavelengths of light travel at different speeds through an optical fiber. To minimize this particular dispersion phenomenon, two factors must be managed. The first of these is the number of wavelengths that make up the transmitted signal. An LED, for example, emits a rather broad range of wavelengths between 30 and 180 nm, whereas a laser emits a much narrower spectrum—typically less than 5 nm. Thus, a laser's output is far less prone to be seriously affected by material dispersion than the signal from an LED.

The second factor that affects the degree of material dispersion is a characteristic called the *center-operating wavelength of the source signal.* In the vicinity of 850 nm, red, longer wavelengths travel faster than their shorter blue counterparts, but at 1,550 nm, the situation is the opposite: blue wavelengths travel faster. Of course, a point occurs at which the two meet and

Figure 4-21
Narrow core
optical fiber.

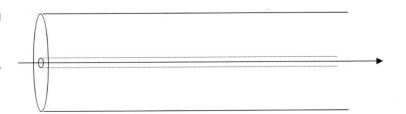

share a common minimum dispersion level; it is in the range of 1,310 nm, often referred to as the *zero dispersion wavelength*. Clearly, this is an ideal place to transmit data signals because dispersion effects are minimized here. As we will see later, however, other factors crop up that make this a less desirable transmission window than it appears. Material dispersion is a particularly vexing problem in single-mode fibers.

Waveguide Dispersion Because the core and the cladding of a fiber have slightly different indices of refraction, the light that travels in the core moves slightly slower than the light that escapes into and travels in the cladding. This results in a dispersion effect that can be corrected by transmitting at specific wavelengths where material and waveguide dispersion actually occur at minimums.

Putting It All Together

So what does all of this have to do with the high-speed transmission of voice, video, and data? A lot, as it turns out. Understanding where attenuation and dispersion problems occur helps optical design engineers determine the best wavelengths at which to transmit information, taking into account distance, type of fiber, and other factors that can potentially affect the integrity of the transmitted signal. Consider the graph shown in Figure 4-22. It depicts the optical transmission domain, as well as the areas where problems arise. Attenuation (dB/km) is shown on the Y-Axis; wavelength (nm) is shown on the X-Axis.

First of all, note that ofour *transmission windows* are in the diagram. The first one is at approximately 850 nm, the second at 1,310 nm, the third at 1,550 nm, and the fourth at 1,625 nm; the last two are labeled the C- and L-band, respectively. The 850 nm band was the first to be used because of its adherence to the wavelength at which the original LED technology operated. The second window at 1,310 nm enjoys low dispersion; this is where dispersion effects are minimized. 1,550 nm, the so-called C-Band, has emerged as the ideal wavelength at which to operate long-haul systems and systems upon which DWDM has been deployed because loss is minimized in this region and dispersion minimums can be shifted here. The relatively new L-Band has enjoyed some early success as the next effective operating window. A new band, the S-Band, is currently under development.

Notice also that Rayleigh Scattering is shown to occur at or around 1,000 nm, whereas hydroxyl absorption by water occurs at 1,240 and 1,390 nm.

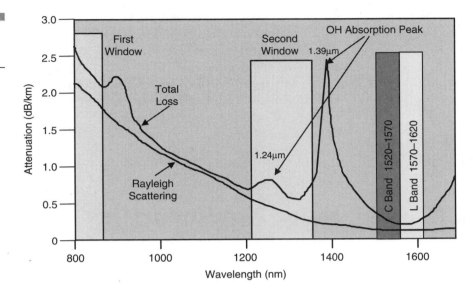

Figure 4-22
Optical transmission domain.

Needless to say, network designers would be well served to avoid transmitting at any of the points on the graph where Rayleigh Scattering, high degrees of loss, or hydroxyl absorption have the greatest degree of impact. Notice also that dispersion, shown by the lower line, is at a minimum point in the second window, whereas loss, shown by the upper line, drops to a minimum point in the third window. In fact, dispersion is minimized in traditional single-mode fiber at 1,310 nm, whereas loss is at a minimum at 1,550 nm. So the obvious question becomes this: which one do you want to minimize—loss or dispersion?

Luckily, this choice no longer has to be made. Today, *dispersion-shifted fibers* (DSF) have become common. By modifying the manufacturing process, engineers can shift the point at which minimum dispersion occurs from 1,310 nm to 1,550 nm, causing it to coincide with the minimum loss point such that loss and dispersion occur at the same wavelength.

Unfortunately, although this fixed one problem, it created a new and potentially serious alternative problem. Dense Wavelength Division Multiplexing (DWDM) has become a mainstay technology for multiplying the available bandwidth in optical systems. When DWDM is deployed over dispersion-shifted fiber, serious nonlinearities occur at the zero dispersion point, which effectively destroy the DWDM signal. Think about it: DWDM relies on the capability to channelize the available bandwidth of the optical infrastructure and maintain some degree of separation between the channels. If dispersion is minimized in the 1,550-nm window, then the channels will effectively overlay each other in DWDM systems. Specifically, a problem called *four-wave mixing* creates *sidebands* that interfere with the DWDM channels, destroying their integrity. In response, fiber manufacturers have created *non-zero dispersion-shifted fiber* (NZDSF) that lowers the dispersion point to near zero and makes it occur just outside of the 1,550-nm window. This eliminates the nonlinear four-wave mixing problem.

Fiber Nonlinearities

A classic business quote, imminently applicable to the optical networking world, observes "in its success lie the seeds of its own destruction." As the marketplace clamors for longer transmission distances with minimal amplification, more wavelengths per fiber, higher bit rates, and increased signal power, a rather ugly collection of transmission impairments, known as *fiber nonlinearities*, rises to challenge attempts to make them happen. These impairments go far beyond the simple concerns brought about by loss and dispersion; they represent a significant performance barrier.

The Power/Refractive Index Problem

Two fundamental issues result in the bulk of these nonlinearities. The first (and perhaps most critical) is the fact that the refractive index of the core of an optical fiber is directly dependent upon the power of the optical signal that is being transmitted through it. The stronger the transmitted signal, the greater the power-dependent impairment. Because of this relationship, two actions can be taken to minimize the power-related problem. The first is to minimize the transmitted power of the signal, an action that will clearly reduce the impact of power-dependent signal degradation. This, however, has the downside of limiting the transmission distance and is a less-than-desirable option because lower power means that more amplifiers will be required over the long haul. Amplifiers also introduce other problems that are equally annoying. The second solution, which is in some ways more acceptable, is to maximize what is known as the fiber's effective area, a measure of the cross-sectional area of the fiber core that carries the transmitted signal. By broadening the effective area of the fiber, it has the capability to gather more of the transmitted signal and reduce the need for an inordinately strong signal. Lucent's TrueWave® Fiber and Corning's *Large Effective Area Fiber* (LEAF®) are examples of specially engineered products designed to overcome this problem.

The special relationship that exists between transmission power and the refractive index of the medium gives rise to four service-affecting optical nonlinearities: *self-phase modulation* (SPM), *cross-phase modulation* (XPM), *four-wave mixing* (FWM), and *intermodulation*.

Self-Phase Modulation (SPM)   When self-phase modulation occurs, chromatic dispersion kicks in to create something of a technological double whammy. As the light pulse moves down the fiber, its leading edge increases the refractive index of the core, causing a shift toward the blue end of the spectrum. The trailing edge, on the other hand, decreases the refractive index of the core, causing a shift toward the red end of the spectrum. This causes an overall spreading or smearing of the transmitted signal, a phenomenon known as *chirp*. It occurs in fiber systems that transmit a single pulse down the fiber and is proportional to the amount of chromatic dispersion in the fiber: the more chromatic dispersion, the more SPM. It is counteracted with the use of large effective area fibers.

Cross-Phase Modulation (XPM)   When multiple optical signals travel down the same fiber core, they both change the refractive index in direct proportion to their individual power levels. If the signals happen to cross, they will distort each other (remember as Egon warned in the movie *Ghostbusters*, "Don't cross the streams!"). Although XPM is similar to SPM, it has one significant difference: whereas self-phase modulation is directly affected by chromatic dispersion, cross-phase modulation is only minimally affected by it. Large effective area fibers can reduce the impact of XPM.

Four-Wave Mixing (FWM)   Four-wave mixing is the most serious of the power/refractive index-induced nonlinearities today because it has a catastrophic effect on DWDM-enhanced systems. Because the refractive index of fiber is nonlinear and because multiple optical signals travel down the fiber in DWDM systems, a phenomenon known as *third-order distortion* can occur that seriously affects multichannel transmission systems. Third-order distortion causes harmonics to be created in large numbers that have the annoying habit of occurring where the actual signals are, resulting in their obliteration. These harmonics tend to become numerous according to the equation,

$$1/2(N^3 - N^2)$$

where N is the number of signals. So if a DWDM system is transporting 16 channels, the total number of potentially destructive harmonics created would be 1,920.

Four-wave mixing is directly related to DWDM. In DWDM fiber systems, multiple simultaneous optical signals are transmitted across an optical span. They are separated on an ITU-blessed standard transmission grid by as much as 100 GHz (although most manufacturers today have reduced that to 50 GHz or better). This separation ensures that they do not interfere with each other.

Consider now the effect of dispersion-shifted fiber on DWDM systems. In DSF, signal transmission is moved to the 1,550-nm band to ensure that dispersion and loss are both minimized within the same window. However, minimal dispersion has a rather severe, unintended consequence when it occurs in concert with DWDM: because it reduces dispersion to near zero, it also prevents multichannel systems from existing because it does not provide proper channel spacing. Four-wave mixing, then, becomes a serious problem.

Several things can reduce the impact of FWM. As the dispersion in the fiber drops, the degree of four-wave mixing increases dramatically. In fact, it is worst at the zero dispersion point. Thus, the intentional inclusion of a small amount of chromatic dispersion actually helps to reduce the effects of FWM. For this reason, fiber manufacturers sell non-zero dispersion-shifted fiber, which moves the dispersion point to a point near the zero point, thus ensuring that a small amount of dispersion creeps in to protect against FWM problems.

Another factor that can minimize the impact of FWM is to widen the spacing between DWDM channels. This, of course, reduces the efficiency of the fiber by reducing the total number of available channels and is therefore not a popular solution, particularly because the trend in the industry is to move toward narrower channel spacing as a way to increase the total number of available channels. Already several vendors have announced spacing as narrow as 5 GHz. Finally, large effective area fibers tend to suffer less from the effects of FWM.

Intermodulation Effects   In the same way that cross-phase modulation results from interference between multiple simultaneous signals, intermodulation causes secondary frequencies to be created that are cross-products of the original signals being transmitted. Large effective area fibers can alleviate the symptoms of intermodulation.

Scattering Problems

Scattering within the silica matrix causes the second major impairment phenomenon. Two significant nonlinearities result: *Stimulated Brillouin Scattering* (SBS) and *Stimulated Raman Scattering* (SRS).

Stimulated Brillouin Scattering (SBS)   SBS is a power-related phenomenon. As long as the power level of a transmitted optical signal remains below a certain threshold, usually on the order of 3 milliwatts, SBS is not a problem. The threshold is directly proportional to the fiber's effective area, and because dispersion-shifted fibers typically have smaller effective areas, they have lower thresholds. The threshold is also proportional to the width of the originating laser pulse: as the pulse gets wider, the threshold goes up. Thus, steps are often taken through a variety of techniques to artificially broaden the laser pulse. This can raise the threshold significantly, to as high as 40 milliwatts.

SBS is caused by the interaction of the optical signal moving down the fiber with the acoustic vibration of the silica matrix that makes up the fiber. As the silica matrix resonates, it causes some of the signal to be reflected back toward the source of the signal, resulting in noise, signal degradation, and a reduction of overall bit rate in the system. As the power of the signal increases beyond the threshold, more of the signal is reflected, resulting in a multiplication of the initial problem.

It is interesting to note that two forms of Brillouin Scattering are actually available. When (sorry, a little more physics) electric fields that oscillate in time within an optical fiber interact with the natural acoustic resonance of the fiber material itself, the result is a tendency to backscatter light as it passes through the material; this is called *Brillouin Scattering*. If, however, the electric fields are caused by the optical signal itself, the signal is seen to cause the phenomenon; this is called *Stimulated Brillouin Scattering*.

To summarize: because of backscattering, SBS reduces the amount of light that actually reaches the receiver and causes noise impairments. The problem increases quickly above the threshold and has a more deleterious impact on longer wavelengths of light. One additional fact: in-line optical amplifiers such as *erbium-doped fiber amplifiers* (EDFAs) add to the problem significantly. If four optical amplifiers are present along an optical span, the threshold will drop by a factor of four. Solutions to SBS include the use of wider-pulse lasers and larger effective area fibers.

Stimulated Raman Scattering (SRS)   SRS is something of a power-based crosstalk problem. In SRS, high-power, short-wavelength channels tend to bleed power into longer-wavelength, lower-power channels. It occurs when a light pulse moving down the fiber interacts with the crystalline matrix of the silica, causing the light to be backscattered and shift the wavelength of the pulse slightly. Whereas SBS is a backward-scattering phenomenon, SRS is a two-way phenomenon, causing both backscattering and a wavelength shift. The result is crosstalk between adjacent channels.

The good news is that SRS occurs at a much higher power level—close to a watt. Furthermore, it can be effectively reduced through the use of large effective area fibers.

An Aside: Optical Amplification

As long as we are on the subject of Raman Scattering, we should introduce the concept of optical amplification. This may seem like a bit of a non sequitur, but it really isn't; true optical amplification actually uses a form of Raman Scattering to amplify the transmitted signal!

Traditional Amplification and Regeneration Techniques

In a traditional metallic analog environment, transmitted signals tend to weaken over distance. To overcome this problem, amplifiers are placed in the circuit periodically to raise the power level of the signal. This technique has a problem, however: in addition to amplifying the signal, amplifiers also amplify whatever cumulative noise has been picked up by the signal during its trip across the network. Over time, it becomes difficult for a receiver to discriminate between the actual signal and the noise embedded in the signal. Extraordinarily complex recovery mechanisms are required to discriminate between optical wheat and noise chaff.

In digital systems, *regenerators* are used to not only amplify the signal, but to also remove any extraneous noise that has been picked up along the way. Thus, digital regeneration is a far more effective signal recovery methodology than simple amplification.

Even though signals propagate significantly farther in optical fiber than they do in copper facilities, they are still eventually attenuated to the point that they must be regenerated. In a traditional installation, the optical signal is received by a receiver circuit, converted to its electrical analog, regenerated, converted back to an optical signal, and transmitted onward over the next fiber segment. This *optical-to-electrical-to-optical* (O-E-O) conversion process is costly, complex, and time consuming. However, it is proving to be far less necessary as an amplification technique than it used to be because of true optical amplification that has recently become commercially feasible. Please note that optical amplifiers *do not* regenerate signals; they merely amplify. Regenerators are still required, albeit far less frequently.

Optical amplifiers represent one of the technological leading edges of data networking. Instead of the O-E-O process, shown in Figure 4-23, optical amplifiers receive the optical signal, amplify it as an optical signal, and then retransmit it as an optical signal—no electrical conversion is required, as shown in Figure 4-24. Like their electrical counterparts, however, they also amplify the noise; at some point, signal regeneration is required.

Optical Amplifiers: How They Work

It was only a matter of time before all-optical amplifiers became a reality. It makes intuitively clear sense that a solution that eliminates the electrical portion of the O-E-O process would be a good one; optical amplification is that solution.

You will recall that SRS is a fiber nonlinearity that is characterized by high-energy channels pumping power into low-energy channels. What if

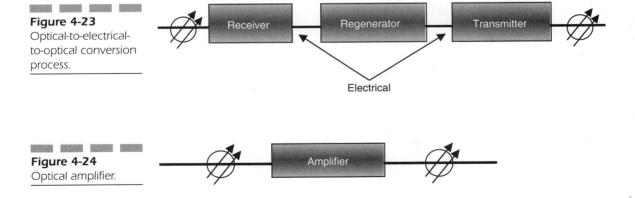

Figure 4-23
Optical-to-electrical-to-optical conversion process.

Figure 4-24
Optical amplifier.

that phenomenon could be harnessed as a way to amplify optical signals that have weakened over distance?

Optical amplifiers are actually rather simple devices that as a result tend to be extremely reliable. As Figure 4-25 illustrates, the optical amplifier comprises the following: an input fiber, carrying the weakened signal that is to be amplified; a pair of optical isolators; a coil of doped fiber; a pump laser; and the output fiber that now carries the amplified signal.

The coil of doped fiber lies at the heart of the optical amplifier's functionality. Doping is simply the process of embedding some kind of functional impurity in the silica matrix of the fiber when it is manufactured. In optical amplifiers, this impurity is more often than not an element called *erbium*. Its role will become clear in just a moment.

The pump laser shown in the upper-left corner of Figure 4-25 generates a light signal at a particular frequency—often times 980 nm—in the opposite direction that the actual transmitted signal flows. As it turns out, erbium becomes atomically excited when it is struck by light at that wavelength. When an atom is excited by pumped energy, it jumps to a higher energy level (those of you who are recovering physicists will remember classroom discussions about orbital levels—$1S^1$, $1S^2$, $2S^1$, $2S^2$, $2P^6$, and so on), then falls back down, during which it gives off a photon at a certain wavelength. When erbium is excited by light at 980 nm, it emits photons within the 1,550-nm region—coincidentally the wavelength at which multi-

Figure 4-25
Erbium-doped fiber
amplifier (EDFA).

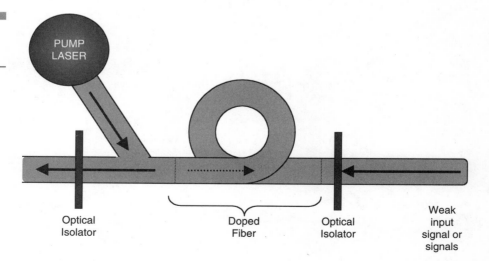

PUMP
LASER

Optical
Isolator

Doped
Fiber

Optical
Isolator

Weak
input
signal or
signals

channel optical systems operate. So when the weak, transmitted signal reaches the coil of erbium-doped fiber, the erbium atoms, now excited by the energy from the pump laser, bleed power into the weak signal at precisely the right wavelength, causing a generalized amplification of the transmitted signal. The optical isolators serve to prevent errant light from backscattering into the system, creating noise.

Of course, EDFAs are highly proletarian in nature: they amplify anything, including the noise that the signal may have picked up. The need for regeneration at some point along the path of long-haul systems will still be present, although far less frequently than in traditional copper systems. Most manufacturers of optical systems publish recommended span engineering specifications that help service providers and network designers take such concerns into account as they design each transmission facility.

Other Amplification Options

At least two other amplification techniques are available in addition to EDFAs that have recently come into favor. The first of these is called *Raman amplification,* which is similar to EDFA in the sense that it relies on Raman effects to do its task, but different for other rather substantial reasons. In Raman amplification, the signal beam travels down the fiber alongside a rather powerful pump beam, which excites atoms in the silica matrix that in turn emit photons that amplify the signal. The advantage of Raman amplification is that it requires no special doping: erbium is not necessary. Instead, the silica itself gives off the necessary amplification. In this case, the fiber itself becomes the amplifier!

Raman amplifiers require a significantly high-power pump beam (about a watt, although some systems have been able to reduce the required power to 750 milliwatts or less) and even at high levels the power gain is relatively low. Their advantage, however, is that their induced gain is distributed across the entire optical span. Furthermore, it will operate within a relatively wide range of wavelengths, including 1,310 and 1,550 nm, currently the two most popular and effective transmission windows.

Semiconductor lasers have also been deployed as optical amplification devices in some installations. In semiconductor optical amplifiers, the weakened optical signal is pumped into the ingress edge of a semiconductor optical amplifier, as shown in Figure 4-26. The active layer of the semiconductor substrate amplifies the signal and regenerates it on the other side. The primary downside to these devices is their size; they are small and their light-collecting capability is therefore somewhat limited. A typical single-mode fiber generates an intense spot of light that is roughly 10 microns in diameter; the point upon which that light impinges upon the semiconductor amplifier is less than a micron in diameter, meaning that much of the light is lost. Other problems also crop up including polarization issues, reflection, and variable gain. As a result, these devices are not in widespread use; EDFAs and Raman amplification techniques are far more common.

Pulling it All Together

So why do we care about these challenges, problems, and nonlinearities? Because they have a direct effect on the degree to which we can transmit signals through optical media. Without question, fiber is orders of magnitude better than copper as a transmission medium for broadband signals, but it does have limitations that cannot be ignored. In the same way that data transmitted over copper networks suffers impairments from cumulative noise and signal deterioration, so too do optical signals. As the demands for higher bandwidth and greater channel counts grow, these impairments must be carefully managed to prevent them from having an inordinately deleterious impact on the systems in which they occur. The good news is that they are well understood, and optical network engineers have developed good measurement tools to detect them and engineering guidelines to control them.

Optical Receivers

So far, we have discussed the sources of light, including LEDs and laser diodes; we have briefly described the various flavors of optical fiber and the

Figure 4-26
Semiconductor laser amplifier.

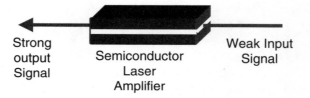

Strong output Signal

Semiconductor Laser Amplifier

Weak Input Signal

problems they encounter as transmission media; now, we turn our attention to the devices that receive the transmitted signal.

The receive devices used in optical networks have a single responsibility: to capture the transmitted optical signal and convert it into an electrical signal that can then be processed by the end equipment. Various stages of amplification may also occur to ensure that the signal is strong enough to be acted upon, and demodulation circuitry may be present, which recreates the originally transmitted electronic signal.

Think for a moment about the term *semiconductor*. A semiconductor is a compound that, well, only semiconducts. It sits in the gray area between conductors and insulators, and must be somehow induced to conduct current. The optical receivers that are commonly used in optical networks are semiconductors themselves; let's take a moment to describe how they work. Sorry, we must descend once again into the depths of physics to do this.

Photosensitive semiconductors, which are silicon-based, typically consist of three functional layers (Figure 4-27): a negative region, a positive region, and a junction region. Photodetectors are said to be *reverse-biased* because the negative charges (electrons) and the positive charges (holes) are not allowed to migrate into the center junction region, thus preventing the flow

Figure 4-27
Three-layer photosensitive semiconductors.

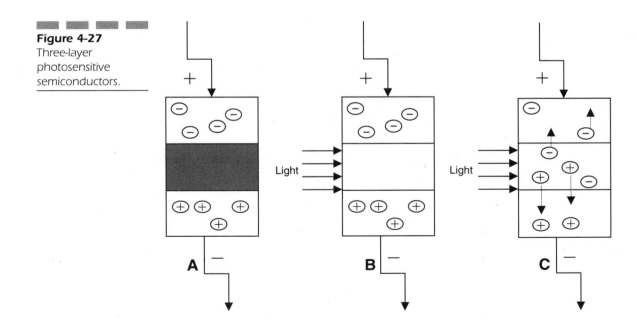

of current through the semiconductor from one active layer to the other (A). This changes (B) when light of a specific wavelength strikes the photosensitive junction layer, creating electron-hole pairs in the junction layer. This results in an overall flow of current that is proportional to the intensity of the light striking the junction layer.

Different substances can be used as photodetectors, including *silicon* (Si), *germanium* (Ge), *gallium arsenide* (GaAs), and *indium-gallium arsenide* (InGaAs), to name a few. They are selected based upon the operating wavelength in which they will be used because their sensitivity to light varies according to the information contained in Figure 4-28.

Photodetector Types

Although many different types of photosensitive devices are available, two types are used most commonly as photodetectors in modern networks: *positive-intrinsic-negative* (PIN) photodiodes and *avalanche photodiodes* (APDs).

Positive-Intrinsic-Negative (PIN) Photodiodes   PIN photodiodes are similar to the device described previously in the general discussion of photosensitive semiconductors. Reverse biasing the junction region of the device prevents current flow until light at a specific wavelength strikes the substance, creating electron-hole pairs and enabling current to flow across the three-layer interface in proportion to the intensity of the incident light. Although they are not the most sensitive devices available for the purpose of photodetection, they are perfectly adequate for the requirements of most optical systems. In cases where they are not considered sensitive enough for high-performance systems, they can be coupled with a preamplifier to increase the overall sensitivity. Figure 4-29 shows an example of a photodetector.

Figure 4-28
Typical
semiconductor
operating
wavelengths.

Substance	Operating Wavelength (nm)
silicon	400–1100
germanium	800–1600
gallium arsenide	400–1000
Indium-gallium arsenide	400–1700

Avalanche Photodiodes (APDs) APDs work as optical signal amplifiers. They use a strong electric field to perform what is known as *avalanche multiplication*. In an APD, the electric field causes current accelerations such that the atoms in the semiconductor matrix get excited and create, in effect, an *avalanche* of current to occur. The good news is that the amplification effect can be as much as 30 to 100 times the original signal; the bad news is that the effect is not altogether linear and can create noise. APDs are sensitive to temperature and require a significant voltage to operate them—30 to 300 volts depending on the device. However, they are popular for broadband systems and work well in the gigabit range.

We have now discussed transmitters, fiber media, and receivers. In the next section, we examine the fibers themselves, and how they have been carefully designed to serve as solutions for a wide variety of networking challenges and to forestall the impact of the nonlinearities described in this section.

Optical Fiber

As was mentioned briefly in a prior section, fiber has evolved over the years in a variety of ways to accommodate both the changing requirements of the customer community and the technological challenges that emerged as the demand for bandwidth climbed precipitously. These changes came in various forms of fiber that presented different behavior characteristics to the market.

Figure 4-29
Photodetector.

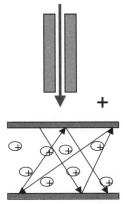

Multimode Fiber

The first of these was multimode fiber, which arrived in a variety of different forms. Multimode fiber bears this name because it enables more than a single mode or ray of light to be carried through the fiber simultaneously because of the relatively wide core diameter that characterizes the fiber (see Figures 4-30 and 4-31). Although the dispersion that potentially results from this phenomenon can be a problem, the use of multimode fiber has advantages. For one thing, it is far easier to couple the relatively wide and forgiving end of a multimode fiber to a light source than that of the much narrower single-mode fiber. It is also significantly less expensive to manufacture (and purchase), and relies on LEDs and inexpensive receivers rather than the more expensive laser diodes and ultrasensitive receiver devices. However, advancements in technology have caused the use of multimode fiber to fall out of favor; single mode is far more commonly used today.

Multimode fiber is manufactured in two forms: *step-index fiber* and *graded-index fiber*. We will examine each in the following sections.

Multimode Step-Index Fiber   In step-index fiber, the index of refraction of the core is slightly higher than the index of refraction of the cladding, as

Figure 4-30
Cross-section of multimode fiber.

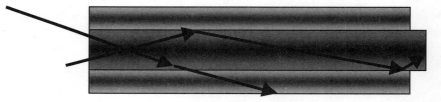

Core index of refraction: 1.5
Cladding index of refraction: 1.3

Figure 4-31
End view showing relative diameter of core in multimode fiber.

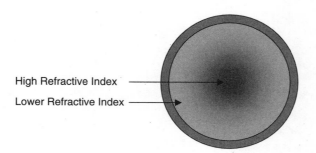

High Refractive Index

Lower Refractive Index

shown in Figure 4-30. Remember that the higher the refractive index, the slower the signal travels through the medium. Thus, in step-index fiber, any light that escapes into the cladding because it enters the core at too oblique an angle will actually travel slightly faster in the cladding (assuming it does not escape altogether) than it would if it traveled in the core. Of course, any rays that are reflected repeatedly as they traverse the core also take longer to reach the receiver, resulting in a dispersed signal that causes problems for the receiver at the other end. Clearly, this phenomenon is undesirable; for that reason, graded-index fiber was developed.

Multimode Graded-Index Fiber   Because of the dispersion that is inherent in the use of step-index fiber, optical engineers created graded-index fiber as a way to overcome the signal degradation that occurred.

In graded-index fiber, the refractive index of the core actually decreases from the center of the fiber outward, as shown in Figure 4-31. In other words, the refractive index at the center of the core is higher than the refractive index at the edge of the core. The result of this rather clever design is that as light enters the core at multiple angles and travels from the center of the core outward, it is actually accelerated at the edge and slowed down near the center, causing most of the light to arrive at roughly the same time. Thus, graded-index fiber helps to overcome the dispersion problems associated with step-index multimode fiber. Light that enters this type of fiber does not travel in a straight line, but rather follows a parabolic path (Figure 4-32), with all rays arriving at the receiver at more or less the same time.

Graded-index fiber typically has a core diameter of 50 to 62.5 microns, with a cladding diameter of 125 microns. Some variations exist; at least one form of multimode graded-index is available with a core diameter of 85 microns, somewhat larger than those described previously. Furthermore, the actual thickness of the cladding is important: if it is thinner than 20 microns, light begins to seep out, causing additional problems for signal propagation.

Figure 4-32
Light propagation in multimode graded index fiber.

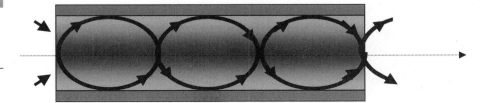

Graded-index fiber was commonly used in telecommunications applications until the late 1980s. Even though graded-index fiber is significantly better than step-index fiber, it is still multimode fiber and does not eliminate the problems inherent in being multimode. Thus, the next generation of optical fiber was born: single-mode fiber.

Modes: An Analogy

The concept of modes is sometimes difficult to understand, so let me pass along an analogy that will help. Imagine a shopping mall that has a wide, open central area that all the shops open onto. An announcement comes over the PA system informing people that "The mall is now closed; please make your way to the exit." Shoppers begin to make their way to the doors on the left, as shown in Figure 4-33, but some wander from store to store, window-shopping along the way, while others take a relatively straight route to the exit. The result is that some shoppers take longer than others to exit the mall because different modes of exiting are available.

Now consider a mall that has a single, very narrow corridor that is only as wide as a person's shoulders. Now when the announcement comes, everyone heads for the exit, but they must form a single file line and head out in an orderly fashion, as shown in Figure 4-34. If you understand the difference between these two examples, you understand single versus multimode fiber. The first example represents multimode; the second represents single mode.

Figure 4-33
"Multimode shoppers" in a shopping mall. Note the various paths that shoppers take as they exit the building.

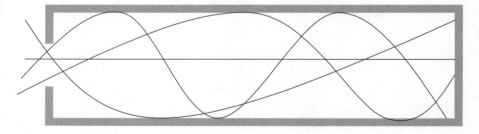

Figure 4-34
"Single mode shoppers." All take the same path as they exit the building.

Single-Mode Fiber

An interesting mental conundrum crops up with the introduction of single-mode fiber. The core of single-mode fiber is significantly narrower than the core of multimode fiber. Because it is narrower, it would seem that its capability to carry information would be reduced due to limited light-gathering capability. This, of course, is not the case. As its name implies, it enables a single mode or ray of light to propagate down the fiber core, thus eliminating the intermodal dispersion problems that plague multimode fibers. In reality, single-mode fiber is a stepped-index design because the core's refractive index is slightly higher than that of the cladding. It has become the de facto standard for optical transmission systems and takes on many forms depending on the specific application within which it will be used.

Most single-mode fiber has an extremely narrow core diameter on the order of 7 to 9 microns, and a cladding diameter of 125 microns. The advantage of this design is that it only allows a single mode to propagate; the downside, however, is the difficulty involved in working with it. The core must be coupled directly to the light source and the receiver in order to make the system as effective as possible; given that the core is approximately one-sixth the diameter of a human hair, the mechanical process through which this coupling takes place becomes Herculean.

Single-Mode Fiber Designs   The reader will recall that we spent a considerable amount of time discussing the many different forms of transmission impairments (nonlinearities) that challenge optical systems. Loss and dispersion are the key contributing factors in most cases, and do in fact cause serious problems in high-speed systems. The good news is that optical engineers have done yeomen's work creating a wide variety of single-mode fibers that address most of the nonlinearities.

Since its introduction in the early 1980s, single-mode fiber has undergone a series of evolutionary phases in concert with the changing demands of the bandwidth marketplace. The first variety of single-mode fiber to enter the market was called *non-dispersion-shifted fiber* (NDSF). Designed to operate in the 1,310-nm second window, dispersion in these fibers was close to zero at that wavelength. As a result, it offered high bandwidth and low dispersion. Unfortunately, it was soon the victim of its own success. As demand for high-bandwidth transport grew, a third window was created at 1,550 nm for single-mode fiber transmission. It provided attenuation levels that were less than half those measured at 1,310 nm, but unfortunately was plagued with significant dispersion. Because the bulk of all installed fiber was NDSF, the only solution available to transmission designers was to narrow the linewidth of the lasers employed in these systems and to make them more powerful. Unfortunately, increasing the power and reducing the laser linewidth is expensive, so another solution soon emerged.

Dispersion-Shifted Fiber (DSF)

One solution that emerged was (DSF. With DSF, the minimum dispersion point is mechanically shifted from 1,310 nm to 1,550 nm by modifying the design of the actual fiber so that waveguide dispersion is increased. The reader will recall that waveguide dispersion is a form of chromatic dispersion that occurs because the light travels at different speeds in the core and cladding.

One technique for building DSF (sometimes called *zero dispersion-shifted fiber*) is to actually build a fiber of multiple layers, as shown in Figure 4-35. In this design, the core has the highest index of refraction and changes gradually from the center outward until it equals the refractive index of the outer cladding. The inner core is surrounded by an inner-cladding layer, which is in turn surrounded by an outer core. This design works well for single-wavelength systems, but experiences serious signal degradation when multiple wavelengths are transmitted, such as when used with DWDM systems. Four-wave mixing, described earlier, becomes a serious impediment to clean transmission in these systems. Given that multiple-wavelength systems are fast becoming the norm today, the single-wavelength limit is a showstopper. The result was a relatively simple and elegant set of solutions.

Figure 4-35
Multilayer zero dispersion-shifted fiber.

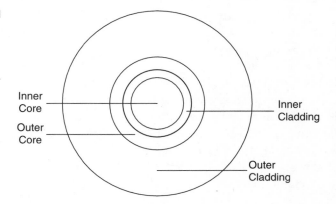

The first of these was to maximize the effective area of the fiber, as was discussed earlier. Lucent's TrueWave® Fiber and Corning's LEAF® Fiber are examples of this. Because the overall power of the optical signal(s) being carried by the fiber is distributed across a broader cross-section, the non-linear performance problems are less pronounced.

The second technique was to eliminate or at least substantially reduce the absorption peaks in the fiber performance graph so that the second and third transmission windows merge into a single larger window, thus enabling the creation of the fourth window described earlier, which operates between 1,565 and 1,625 nm—the so-called L-Band.

Finally, the third solution came with the development of NZDSF which shifts the minimum dispersion point so that it is close to the zero point, but not actually at it. This prevents the nonlinear problems that occur at the zero point to be avoided because it introduces a small amount of chromatic dispersion.

Why Do We Care?

It is always good to go back and review why we care about such things as dispersion-shifting and absorption issues. Remember that the key to keeping the cost of network down is to reduce maintenance and the need to add hardware or additional fiber when bandwidth gets tight. DWDM, discussed in detail later, offers an elegant and relatively simple solution to the problem of the cost of bandwidth. However, its use is not without cost. Multi-wavelength systems will not operate effectively over DSF because of dramatic nonlinearities, so if DWDM is to be used, NZDSF must be deployed.

Summary

In this section, we have examined the history of optical technology and the technology itself, focusing on the three key components within an optical network: the light emitter, the transport medium, and the receiver. We also discussed the various forms of transmission impairment that can occur in optical systems and the steps that have been taken to overcome them.

The result of all this is that optical fiber, once heralded as a near-technological miracle because it only lost 99 percent of its signal strength when transmitted over an entire kilometer, has become the standard medium for the transmission of high-bandwidth signals over great distances. Optical amplification now serves as an augmentation to traditional regenerated systems, enabling the elimination of the optical-to-electrical conversion that must take place in copper systems. The result of all this is an extremely efficient transmission system that has the capability to play a role in virtually any network design in existence today.

We will now examine corollary technologies that add capability and richness to optical transmission including DWDM, optical switching, and routing technologies.

Dense Wavelength Division Multiplexing (DWDM)

When SONET and SDH were first introduced, the bandwidth that they made possible was unheard of. The early systems that operated at OC-3/STM-1 levels (155.52 Mbps) provided volumes of bandwidth that were almost unimaginable. As the technology advanced to higher levels, the market followed Say's Law, creating demand for the ever more available volumes of bandwidth. There were limits, however; today, OC-48/STM-16 (2.5 Gbps) is extremely popular, but OC-192/STM-64 (10 Gbps) represent the practical upper limit of SONET's and SDH's transmission capabilities given the limitations of existing time-division multiplexing technology. The alternative is to simply multiply the channel count—that's where WDM comes into play.

WDM is really nothing more than frequency-division multiplexing, albeit at very high frequencies. The ITU has standardized a channel separation grid that centers around 193.1 Thz, ranging from 191.1 THz to 196.5 THz. Channels on the grid are technically separated by 100 GHz, but many industry players today are using 50-GHz separation.

The majority of WDM systems operate in the C-Band (third window, 1,550 nm), which allows for close placement of channels and the reliance on EDFAs to improve signal strength. Older systems, which spaced the channels 200 GHz (1.6 nm) apart, were referred to as WDM systems; the newer systems are referred to as *Dense* WDM systems because of their tighter channel spacing. Modern systems routinely pack 40 10-Gbps channels across a single fiber, for an aggregate bit rate of 400 Gbps.

How DWDM Works

As Figure 4-36 illustrates, a WDM system consists of multiple input lasers, an ingress multiplexer, a transport fiber, an egress multiplexer, and of course, customer-receiving devices. If the system has eight channels such as the one shown in the diagram, it has eight lasers and eight receivers. The channels are separated by 100 GHz to avoid fiber nonlinearities or closer if the system supports the 50-GHz spacing. Each channel, sometimes referred to as a lambda (λ, the Greek letter and universal symbol used to represent wavelength), is individually modulated, and ideally the signal strengths of the channels should be close to one another. Generally speaking, this is not a problem because in DWDM systems, the channels are closely spaced and therefore do not experience significant attenuation variation from channel to channel.

Operators of DWDM-equipped networks face a significant maintenance issue. Consider a 16-channel DWDM system. This system has 16 lasers, one for each channel, which means that the service provider must maintain 16 spare lasers in case of a laser failure. The latest effort underway is the deployment of tunable lasers, which enable the laser to be tuned to any output wavelength, thus reducing the volume of spares that must be maintained and by extension, the cost.

External Cavity Tunable Lasers are the most common form of tunable light source in use today. Figure 4-37 shows a simplified diagram of an external cavity tunable laser. The laser comprises a laser diode and a rotatable diffraction grating. The laser diode emits a range of wavelengths that must be tuned to a much narrower range to accommodate the stringent requirements of closely spaced DWDM channels.

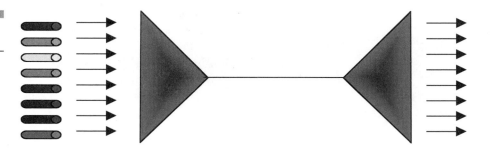

Figure 4-36
Channels in DWDM.

Figure 4-37
External cavity
tunable laser.

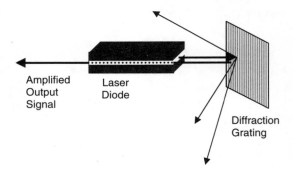

Amplified
Output
Signal

Laser
Diode

Diffraction
Grating

The diffraction grating is an etched array of fine lines, usually photolithographically burned into a segment of fiber by ultraviolet light that can be tuned by rotating it so that it only allows certain wavelengths of light to be reflected directly back to the source—others, as shown in the diagram, are reflected at angles so that they do not enter the laser cavity for amplification and emission. You can simulate the effect of a diffraction grating by holding your index and middle fingers closely together so that a thin space exists between the first and second knuckles of both fingers. Look at a light source through this narrow gap, and by widening and narrowing the gap you will ultimately see the appearance of an array of fine black lines between your fingers. This interference pattern results from the fact that certain wavelengths of visible light are blocked by the narrow gap, in the same way that a diffraction grating prevents certain wavelengths of light from entering the transmission system.

The grating receives the range of emitted wavelengths and reflects all but a narrow range of selected wavelengths away from the laser. The desired wavelengths are reflected back into the laser diode, where they are amplified and emitted from the output facet of the semiconductor. By rotating the filter, different wavelengths can be selected for transmission, thus creating a truly tunable laser.

So what do we find in a typical WDM system? A variety of components, including multiplexers, which combine multiple optical signals for transport across a single fiber; demultiplexers, which disassemble the aggregate signal so that each signal component can be delivered to the appropriate optical receiver (PIN or APD); active or passive switches or routers, which direct each signal component in a variety of directions; filters, which serve to provide wavelength selection; and finally, optical add-drop multiplexers, which give the service provider the ability to pick up and drop off individual wavelength components at intermediate locations throughout the network. Together, these components make up the heart of the typical high-bandwidth optical network. And why is DWDM so important? Because of the cost differential that exists between a DWDM-enhanced network and a traditional network. To expand network capacity today by putting more fiber in the ground costs, on average, about $70K per mile. To add the same bandwidth using DWDM by changing out the endpoint electronics costs roughly one-sixth that amount. Going with the WDM solution clearly has a financial incentive.

Many corporations in the industry manufacture DWDM multiplexers, including Lucent Technologies, Nortel Networks, Ciena, and many others. The next area of focus is switching and routing.

Optical Switching and Routing

DWDM facilitates the transport of massive volumes of data from a source to a destination. Once the data arrives at the destination, however, it must be terminated and redirected to its final destination on a lambda-by-lambda basis. This is done with switching and routing technologies.

Switching versus Routing: What's the Difference?

A review of these two fundamental technologies is probably in order. The two terms are often used interchangeably, and in many cases, a never-ending argument is underway about the differences between the two.

The answer lies in the lower layers of the now-famous OSI Protocol Model, shown in Figure 4-38. OSI is a conceptual model used to study the step-by-step process of transmitting data through a network. It comprises seven layers, the lower three of which define the domain of the typical service provider. These layers, starting with the lowest in the seven-layer stack, are the Physical Layer (Layer 1), the Data Link Layer (Layer 2), and the Network Layer (Layer 3). Layer 1 is responsible for defining the standards and protocols that govern the physical transmission of bits across a medium. SONET and SDH are both Physical Layer standards.

Switching, which lies at Layer 2 (the Data Link Layer) of OSI, is usually responsible for establishing connectivity within a single network. It is a relatively low-intelligence function and is therefore accomplished quite

Figure 4-38
OSI Model layers
and functions.

OSI Model Layers and Functions	
Application Layer	Responsible for payload-specific functions that define the meaning or semantic nuance of the data being transported. Protocols include X.400, and FTAM, EDI.
Presentation Layer	Responsible for general functions that define the form of the transmitted data. Protocols include those related to codeset conversion (EBCDIC, ASCII), compression, and encryption.
Session Layer	Responsible for creating logical linkage between processes running in end systems as well as some security functions.
Transport Layer	Responsible for the end-to-end transmission of the entire message being carried. Protocols include ISO TP, and TCP.
Network Layer	Responsible for control of network congestion and routing functions. Protocols include RIP, OSPF, ARP, IP, and IPX.
Data Link Layer	Responsible for framing the transmitted data unit (usually a packet) so that it can be properly addressed and checked for errors. Protocols include CSMA/CD, ATM, LAPF, LAPD, HDLC, SDLC, token passing, FDMA, TDMA, and CDMA.
Physical Layer	Responsible for the raw transmission of the zeroes and ones that make up the bitstream. Protocols include EIA-232, V.35, and a host of others.

quickly. Such technologies as ATM, frame relay, wireless access technologies such as FDMA, TDMA, and CDMA, and LAN access control protocols (CSMA/CD, token passing) are found at this layer.

Routing, on the other hand, is a Layer 3 (Network Layer) function. It operates at a higher, more complex level of functionality and is therefore more complex. Routing concerns itself with the movement of traffic between sub-networks and therefore complements the efforts of the switching layer. ATM, frame relay, LAN protocols, and the PSTN are switching protocols; IP, RIP, OSPF, and IPX are routing protocols.

Switching in the Optical Domain

The principal form of optical switching is really nothing more than a very sophisticated digital cross-connect system. In the early days of data networking, dedicated facilities were created by manually patching the end points of a circuit at a patch panel, thus creating a complete four-wire circuit. Beginning in the 1980s, digital cross-connect devices such as AT&T's *Digital Access and Cross-Connect* (DACS) became common, replacing the time-consuming, expensive, and error-prone manual process. The digital cross-connect is really a simple switch, designed to establish long-term temporary circuits quickly, accurately, and inexpensively. Figure 4-39 illustrates the cross-connect.

Enter the world of optical networking. Traditional cross-connect systems worked fine in the optical domain, provided no problem occurred going through the O-E-O conversion process. This, however, was one of the aspects of optical networking that network designers wanted to eradicate from their functional requirements. Thus, the optical cross-connect switch was born.

The first of these to arrive on the scene was Lucent Technologies' LambdaRouter. Based on a switching technology called *Micro Electrical Mechanical System* (MEMS), the LambdaRouter was the world's first all-optical cross-connect device.

MEMS relies on micro-mirrors, an array of which is shown in Figure 4-40. The mirrors can be configured at various angles to ensure that an incoming lambda strikes one mirror, reflects off a fixed mirrored surface, strikes another movable mirror, and is then reflected out an egress fiber. The LambdaRouter is now commercially deployed and offers speed, a relatively small footprint, bit rate, and protocol transparency, non-blocking architecture, and highly developed database management. Fundamentally, these devices are very high-speed, high-capacity switches or cross-connect devices. They are not routers because they do not perform Layer 3 functions. All of the major manufacturers, however, including Lucent, Nortel, Ciena, Agilent, and Juniper, have announced initiatives designed to craft true optical routers, and Layer 3 protocols such as MPLS, OSRP, and GMPLS are being tested at the time of this writing which will purportedly add routing functionality to the all-optical domain.

Figure 4-39
Schematic diagram of cross-connect system.

Figure 4-40
MEMS array
(Courtesy Lucent
Technologies).

Years ago when I worked for the telephone company, I was often asked, "how does it all work?" My most common response was, "it's all smoke and mirrors." Well, the mirror part is certainly true.

In our next chapter, we turn our attention to optical switching with a discussion of one critical component, *Asynchronous Transfer Mode* (ATM).

Asynchronous Transfer Mode (ATM)

The Winchester Mystery House in San Jose, California is a 160-room, Victorian mansion designed and built by Sarah Winchester, the wife of William Winchester, son of the manufacturer of the repeating rifle that bears his name. Fifteen years after marrying William, he died of tuberculosis, leaving Sarah a widow.

Upset by the death of her husband, Sarah took the advice of friends and family and consulted spirit guides as a way to help lessen the pain of loss. They convinced Sarah that the spirits of those who had been killed by Winchester rifles haunted her, and that those same spirits were exacting revenge on the Winchester family. Furthermore, they convinced Sarah that those same spirits had placed a curse on her with every intention of haunting her for the rest of her life.

The only saving grace was that Sarah's advisors informed her that she could evade the curse by moving from Illinois to California, buying a house, and beginning an endless process of addition and remodeling, creating a warren of doorways and portals throughout the house that would confuse and mislead the spirits. Furthermore, they informed her that as long as she never interrupted the construction process, she would enjoy immortality. So in 1884, she moved to San Jose, purchased an eight-room farmhouse on a large piece of land and immediately began her endless building project. Figure 5-1 shows some of the house's remarkable details.

Figure 5-1
Amazing facts about the Winchester Mystery House.

Winchester Mystery House

- **Number of rooms:** 160
- **Cost:** $5,500,000
- **Date of Construction:** From 1884 until September 5, 1922
- **Number of stories:** 7, prior to 1906 Earthquake; currently 4
- **Number of acres:** Originally 161.919; currently 4
- **Number of basements:** 2
- **Heating:** Steam, forced air, fireplaces
- **Number of windows:** Approximately 10,000
- **Number of doors:** Approximately 467
- **Number of fireplaces:** 47 (gas, wood, or coal burning)
- **Number of chimneys:** Currently 17 with evidence of 2 others
- **Number of bedrooms:** Approximately 40
- **Number of kitchens:** 5 or perhaps 6
- **Number of staircases:** 40
- **Number of gallons of paint required to paint entire home:** Over 20,000

Sarah never had a plan for the design of her ever-expanding house. She built randomly, and the mansion sprawled over the surrounding landscape, absorbing multiple structures on the 161-acre estate like a gigantic amoeba. She built steadily, 24 hours a day, for 38 years until her death in 1922. The result of her rather strange effort is a cacophony of architecture: the house has French doors that open onto 30-foot drops, staircases that end at the ceiling, hallways that leave a room, turn left twice, and reenter the room, and doorways that open onto lathe and plaster walls.

You may wonder why I chose to begin this discussion of ATM with the story of Sarah Winchester's house. The answer is quite simple: corporate networks often evolve, in the same way that she built her home—with no perceptible long-term plan. They begin with the best intentions, usually a compact collection of homogeneous technologies designed to perform a specific set of services for a limited number of users. Over time, however, as business requirements evolve and access and transport technologies advance, networks become heterogeneous, chaotic, and often remarkably difficult to manage. Meanwhile, network managers find themselves wishing for a single full-service network that can be all things to all applications and services. This was part of the reason why ATM was created.

Network architectures often develop in lockstep with the corporate models that they serve. For example, companies with centralized management authorities such as utilities, banks, and hospitals often have centralized, tightly controlled, hierarchical data-processing environments (illustrated in Figure 5-2) to protect the sensitive data that resides in their massive databases. In these companies, all roads lead to the mainframe computer because it best meets their specific needs. On the other hand, organizations that are distributed in nature, such as professional research and development entities, universities, and film-production companies, often have well-developed, highly distributed data-processing architectures that meet the needs of the distributed functional entities of those businesses. They tend to share information on more of a peer-to-peer basis and their corporate structures reflect this reality. Consider the splash that companies like Hewlett-Packard, Xerox, and Apple made in the early 1980s with announcements about their corporate-managerial models. As a result, their network models reflect this connectivity, thus the ongoing growth of such technologies as Ethernet and other LAN-based peer-to-peer network models.

Today, telecommunications-service providers face competition from a variety of both traditional and non-traditional sources. The characteristics that have typically defined a company's competitive position—usually access to virtually unlimited capital and deployment of the newest technologies—no longer apply. Everyone in the industry has deep pockets today,

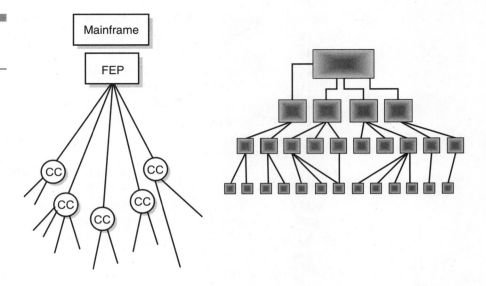

and although the best technology can provide a competitive advantage for a fleeting moment, in today's world of rapid technological advancement, the advantage erodes and disappears quickly. Although we often hear that "time is money," a more applicable follow-on adage might be, "yeah, but speed is profit." The first one to market with a product or service wins, and in today's technology-dependent marketplace, high-speed, multiservice network architectures can mean the difference between success and failure. Furthermore, the ability to bring definable, tangible value to the customer (value defined by the customer, by the way) has an enormous impact on a company's retention of its competitive advantage.

ATM came about not only because of the proliferation of diverse network architectures, but also because of the evolution of traffic characteristics and payload-transport requirements. To the relatively straightforward demands of voice we add high- and low-speed bulk data, streaming and stored video, MP3, imaging, and a variety of other multimedia content types that place increasing demands on the network for service in the form of readily available bandwidth. Furthermore, we have seen a requirement arise for a mechanism that can transparently and correctly transport the *mix* of various traffic types over a single network infrastructure, while at the same time delivering granular, controllable, and measurable *quality of service* (QoS) levels for each service type. In its original form, ATM was designed to do exactly that, working with SONET or SDH to deliver high-speed transport and switching throughout the network—in the wide area, the metropolitan area, the campus environment, and the local area network, right down to the desktop, seamlessly, accurately, and fast.

Today, because of competition from such technologies as QoS-aware IP transport, proprietary high-speed mesh networks, and Fast and Gigabit Ethernet, ATM has, for the most part, lost the race to the desktop. ATM is a cell-based technology, which means that the fundamental unit of transport—a frame of data, if you will—is of a fixed size, which enables switch designers to build faster, simpler devices because they can always count on their switched payload being the same size at all times. That cell comprises a five-octet header and a 48-octet payload field, as shown in Figure 5-3. The payload contains user data; the header contains information that the network requires to both transport the payload correctly and ensure proper quality of service levels for the payload. ATM accomplishes this task well, but at a cost. The five-octet header comprises nearly 10 percent of the cell, a rather significant price to pay, particularly when other technologies such as IP and SONET add their own significant percentages of overhead to the overall payload.

This reality is part of the problem: ATM's original claims to fame and the reasons it rocketed to the top of the technology hit parade came from its cability to switch cells at tremendous speed through the fabric of the wide area network and the ease with which the technology could be scaled to fit any network situation. Today, however, given the availability of high-speed IP routers that routinely route packets at terabit rates, ATM's advantages have begun to pale to a certain degree.

A Technological Phoenix

ATM has, however, emerged from the flames in other ways. Today, many service providers see ATM as an ideal aggregation technology for diverse

Figure 5-3
ATM cell.

Header   Payload

5 octets   48 octets

traffic streams that need to be combined for transport across a wide area network that will most likely be IP-based. ATM devices, then, will be placed at the edge of the network, where they will collect traffic for transport across the Internet or (more likely) a privately-owned IP network. Furthermore, because it has the capability to be something of a chameleon by delivering diverse services across a common network fabric, it is further guaranteed a seat at the technology game.

It is interesting to note that the traditional, legacy telecommunications network comprises two principal regions that can be clearly distinguished from each other: the network itself, which provides switching, signaling, and transport for traffic generated by customer applications; and the access loop, which provides the connectivity between the customer's applications and the network. In this model, the network is considered to be a relatively intelligent medium, whereas the customer equipment is usually considered to be relatively stupid.

Not only is the intelligence seen as being concentrated within the confines of the network, so too is the bulk of the bandwidth because the legacy model indicates that traditional customer applications don't require much of it. However, between central-office switches and between the offices themselves, enormous bandwidth is required.

Today, this model is changing. Customer equipment has become remarkably intelligent, and many of the functions previously done within the network cloud are now performed at the edge. PBXs, computers, and other devices are now capable of making discriminatory decisions about required service levels, eliminating any need for the massive intelligence embedded in the core.

At the same time, the bandwidth is migrating from the core of the network toward the customer as applications evolve to require it. Massive bandwidth is within the cloud, but the margins of the cloud are expanding toward the customer.

The result of this evolution is a redefinition of the network's regions. Instead of a low-speed, low-intelligence access area and a high-speed, high-intelligence core, the intelligence has migrated outward to the margins of the network and the bandwidth. What was once exclusively a core resource is now equally distributed at the edge. Thus, we see something of a core and edge distinction evolving as customer requirements change.

One reason for this steady migration is the well-known fact within sales and marketing circles that products sell best when they are located close to the buying customer. They are also easier to customize for individual customers when they are physically closest to the situation for which the customer is buying them.

In his seminal paper, "The Rise of the Stupid Network," David Isenberg makes the following observation:

The Intelligent Network is a straight-line extension of the four assumptions above—scarcity, voice, circuit switching, and control. Its primary design impetus was not customer service. Rather, the Intelligent Network was a telephone company attempt to engineer vendor independence, more automatic operation, and some "intelligent" new services into existing network architecture. However, even as it rolls out and matures, the Intelligent Network is being superseded by a Stupid Network, with nothing but dumb transport in the middle and intelligent user-controlled endpoints, whose design is guided by plenty, not scarcity, where transport is guided by the needs of the data, not the design assumptions of the network.

Isenberg continues

A new network "philosophy and architecture" is replacing the vision of an Intelligent Network. The vision is one in which the public communications network would be engineered for always-on use, not intermittence and scarcity. It would be engineered for intelligence at the end-user's device, not in the network. And the network would be engineered simply to "Deliver the Bits, Stupid," not for fancy network routing or smart number translation.

Edge vs. Core: What's the Difference?

Edge devices typically operate at the frontier of the network, serving as vital service outposts for their users. Their responsibilities typically include traffic concentration, the process of statistically balancing load against available network resources; discrimination, during which the characteristics of various traffic types are determined; policy enforcement, the process of ensuring that required quality of service levels are available; and protocol internetworking in heterogeneous networks. Edge devices are often the origination point for IP services and typically provide relatively low bandwidth across their backplanes.

Core devices, on the other hand, are responsible for the high-speed forwarding of packet flows from network sources to network destinations. These devices respond to directions from the edge and ensure that resources are available across the wide area network to ensure that quality of service is guaranteed on an end-to-end basis. They tend to be more robust devices than their edge counterparts, and typically offer massive through-put across their backplanes. They are non-blocking and support large numbers of high-speed interfaces.

As the network has evolved to this edge/core dichotomy, the market has evolved as well. As convergence continues to advance, and multiprotocol, multimedia networks become the rule rather than the exception, sales grow apace. By 2003, consultancy RHK estimates that the edge switch and router market will exceed $21 billion, whereas the core market will be nearly $16 billion. In the core, Cisco currently holds the bulk of the market at roughly 50 percent, slightly less at the edge with 31 percent. Other major players include Lucent Technologies, Marconi, Nortel Networks, Juniper, Newbridge, Fore, Avici, and a host of smaller players.

It is interesting to note that in order to adequately implement convergence, the network must undergo a form of divergence as it is redesigned in response to consumer demands. As we just described, the traditional network concentrates its bandwidth and intelligence in the core. The evolving network has in many ways been inverted, moving the intelligence and traffic-handling responsibilities out to the user, replacing them with the high-bandwidth core described earlier. In effect, the network becomes something of a high-tech donut, as shown in Figure 5-4.

Figure 5-4
The divergence donut.

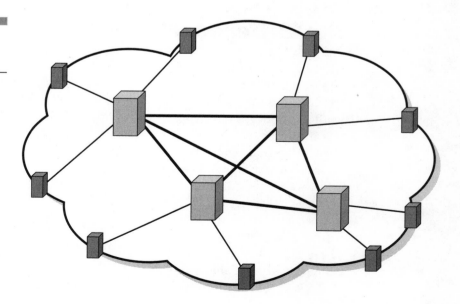

The core, then, becomes the domain of optical networking at its best, offering massively scalable bandwidth through routers capable of handling both high-volume traffic and carrying out the QoS mandates of the edge devices that originate the traffic.

The drivers behind this technology schism are similar to those cited earlier. They include

- The need to form routes on demand between individual users as well as between disparate work groups, in response to the market avoidance of high-cost dedicated facilities
- Guaranteed interoperability between disparate (but widely utilized) protocols
- Universal, seamless connectivity between multiple-corporate locations
- Optimum utilization of network bandwidth through intelligent prioritization and routing techniques
- Traffic aggregation for wide area transport to ensure efficient use of network bandwidth
- Granular quality of service control through effective policy and queue management techniques
- Growing deployment and acceptance of high-speed access technologies such as DSL, cable modems, wireless local-loop solutions, and satellite connectivity

Why is this evolution occurring? Because the closer a service provider places its services to the customer, the more customized, targeted, and immediate those services can be made available to that customer. When they are delivered from a shared central office, they are much more generalized, catering more to the requirements of the masses and treating the customer as if his or her requirements were commodities. As the network evolves and a clear functional delineation between the edge and the core becomes obvious, the role of the central office suddenly changes. In fact, the central office largely disappears. Instead of a massive centralized infrastructure from which all services are delivered (similar to the model employed in legacy data centers), we now see the deployment of hundreds of smaller regional offices placed closer to customer concentrations and housing the edge switching and routing technologies that deliver the discrete services to each customer on an as-required basis. Those smaller offices are in turn connected to the newly evolved optical core, which replaces the legacy central office and delivers high-speed transport for traffic to and from the customers connected to the edge offices. This is the network model of the future: it is more efficient and places both the services and the bandwidth that they require where they belong.

What does this have to do with ATM and SONET? A lot, as it turns out. Let's consider one simple example of the distributed central office that I allude to in the previous paragraph. DSL is a marvelous technology that offers customers the ability to enjoy broadband access to the network, yet its level of penetration remains depressingly low. Part of the reason for this is the fact that most residential local loops in the U.S. are deployed over digital-loop carrier systems, as illustrated in Figure 5-5. This design makes sense for service providers for several reasons. First of all, consider the drawing. In the drawing, 96 telephone customers are provided with voice service over five T-1 carrier facilities, requiring a total of 20 conductors (wires) between the *central office terminal* (COT) and the *remote terminal* (RT). If it were not for the carrier system, those same 96 customers would require 192 conductors, a significant increase and cost. The only problem with this model is that each customer is assigned a single 64-Kbps time slot, which effectively eliminates his or her ability to enjoy higher bandwidth services. To counter this service limitation, some service providers have chosen to move the central office-based *Digital Subscriber Loop Access Multiplexer* (DSLAM) from the CO to the residential neighborhood by hanging it on the end of an optical fiber and using ATM protocol over that fiber to carry out QoS mandates and high-speed switching requirements as they crop up. The customer can then attach directly to the DSLAM, thus

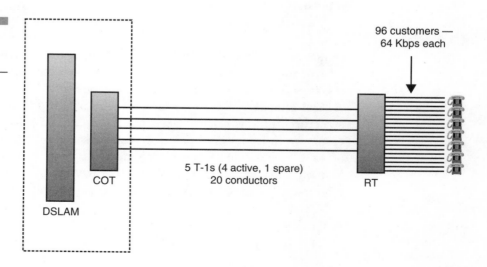

Figure 5-5
A subscriber loop carrier system.

eliminating the legacy bottleneck that occurs with carrier- system service. This is shown in Figure 5-6.

Of course, the network management model must now change in response to the new network architecture. Instead of managing centralized network elements (a relatively simple task) the management system must now manage in a distributed environment. This is more complex, but if done properly results in far better customer service because of the immediacy that results from managing at the customer level.

ATM Technology Overview

Because ATM plays such a major role in SONET networks today, it is important to develop at least a rudimentary understanding of its functions, architectures, and offered services. We begin this discussion of ATM with the overall structure of a typical ATM network, starting from the perspective of the user and working toward the network cloud.

The access interface between the user's equipment and the first ATM switch is called the *User-to-Network Interface* (UNI) (see Figure 5-7). The user's equipment may be something as simple as a DSU, a router behind a DSU, or a private, local ATM switch. If the user's equipment is another

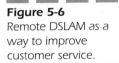

Figure 5-6
Remote DSLAM as a way to improve customer service.

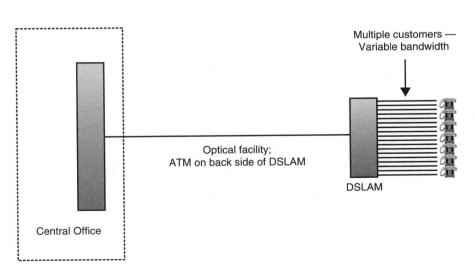

Multiple customers — Variable bandwidth

Optical facility; ATM on back side of DSLAM

DSLAM

Central Office

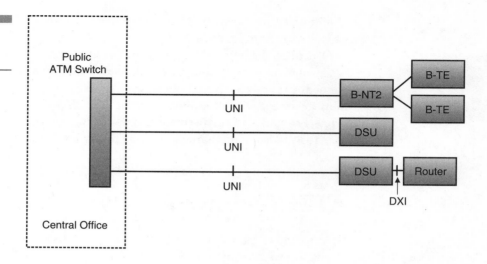

Figure 5-7
The ATM network protocol model.

switch or an ATM-capable PBX, the device is referred to as a Broadband-NT2. Devices that access this local switch are known as *Broadband Terminal Equipment* (B-TEs). In the event that the user's equipment is a router and the router is not capable of interfacing directly to the ATM network because of a protocol incompatibility, then the router will communicate with a DSU over an *ATM Data Exchange* (DXI) Interface. The DSU will in turn communicate with the network. In some cases, for cost and complexity reasons, the router will send its data directly to the ATM switch in the cloud without first converting it to cells. In those cases, the data will be sent to the switch inside a layer-two frame, across what is known as a *Frame-Based UNI* (FUNI).

In the event that the ATM network is being used to emulate LAN services across a wider area, a *LAN Emulation* (LANE) server may be deployed. In this case, the Broadband-TEs will access the server via a *LANE UNI* (LUNI). The LANE server, in turn, will access the local ATM switch (which is probably privately owned) across a *Private Network Node Interface* (PNNI).

Inside the cloud, the ATM switches communicate with one another across a *Network Node Interface* (NNI). Public ATM clouds communicate with each other across a standardized interface in the same way that traditional public packet-switched networks communicate with each other. The *Broadband Interexchange Carrier Interface* (B-ICI) is used to interconnect discrete networks.

These interfaces, particularly those installed to interconnect public and private networks, must implement signaling and routing functions between the two and must therefore rely on the appropriate standards to do so. Organizations such as the ITU-T, the ATM Forum, ANSI, Telcordia, and the IETF have all contributed greatly to the development of standards that address these and other issues.

Addressing and Signaling in ATM

ATM is a connection-oriented, virtual circuit technology, meaning that communication paths are created through the network prior to actually sending traffic. Once established, the ATM cells are routed based upon a virtual circuit address. A virtual circuit is simply a connection that gives the user the appearance of being dedicated to that user, when in point of fact the only thing that is actually dedicated is a time slot. This technique is generically known as *label-based switching* and is accomplished through the use of routing tables in the ATM switches that designate input ports, output ports, input addresses, output addresses, and quality of service parameters required for proper routing and service provisioning. As a result, cells do not contain explicit destination addresses, but rather contain time slot identifiers.

Every virtual circuit address has two components, as shown in Figure 5-8. The first is the *virtual channel* (VC), which is a unidirectional conduit for the transmission of cells between two endpoints. For example, if two parties are conducting a videoconference, they will each have a VC for the transmission of outgoing cells that make up the video portion of the conference.

The second level of the ATM addressing scheme is called a *virtual path* (VP). A VP is a bundle of VCs that have the same endpoints, and that when

Figure 5-8
Virtual paths and
virtual channels
in ATM.

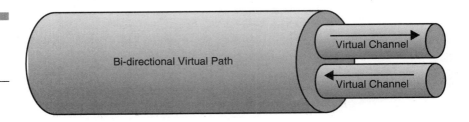

considered together make up a bi-directional transport facility. The combination of unidirectional channels that we need in our two-way videoconferencing example makes up a VP.

ATM Signaling

Signaling in ATM is similar to that used in ISDN and is a subset of the pre-existing ISDN signaling standards. Whereas ISDN uses Q.931 signaling, ATM relies on a subset of Q.931 called Q.2931, which defines signaling packets that establish, maintain, and release VPs and VCs, and negotiate traffic characteristics for each connection at the time that they are established.

The ATM signaling procedures define three types of *signaling virtual channels* (SVCs):

- *Meta-Signaling Virtual Channel (MSVC)*:   The MSVC is a bi-directional permanent channel that is used to create and tear down the temporary SVCs created by users to transfer data. Each interface has one MSVC; the channel is permanent because it must be available so that the switch can signal to the CPE about the status of an incoming call.
- *Signaling Virtual Channel (SVC)*:   This is used for the exchange of actual signaling messages.
- *General Broadcast Signaling Virtual Channel (GBSVC)*:   A channel used by the network to announce an incoming call at all appropriate devices at an interface.

The ATM Cell Header

As we mentioned before, ATM is a cell-based technology that relies on a 48-octet payload field that contains actual user data and a five-byte header that contains information needed by the network to route the cell and provide proper levels of service.

The ATM cell header, shown in Figure 5-9, is examined and updated by each switch it passes through, and comprises six distinct fields: the Generic Flow Control Field, the Virtual Path Identifier, the Virtual Channel Identifier, the Payload Type Identifier, the Cell Loss Priority Field, and the Header Error Control Field, which are described in the following list:

Figure 5-9
ATM cell header.

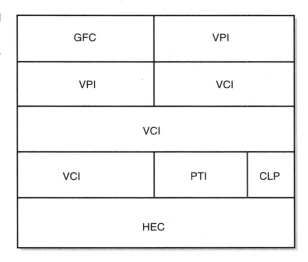

- *Generic Flow Control (GFC)*: This four-bit field is used across the UNI for network-to-user flow control. It has not yet been completely defined in the ATM standards, but some companies have chosen to use it for very specific purposes. For example, Australia's Telstra Corporation uses it for flow control in the network-to-user direction and as a traffic priority indicator in the user-to-network direction.

- *Virtual Path Identifier (VPI)*: The eight-bit VPI identifies the virtual path over which the cells will be routed at the UNI. It should be noted that because of dedicated, internal flow control capabilities within the network, the GFC field is not needed across the NNI. It is therefore redeployed, as shown in Figure 5-10; the four bits are converted to additional VPI bits, thus extending the size of the virtual path field. This provides the identification of more than 4,000 unique VPs. At the UNI, this number is excessive, but across the NNI it is necessary because of the number of potential paths that might exist between the switches that make up the fabric of the network.

- *Virtual Channel Identifier (VCI)*: As the name implies, the 16-bit VCI identifies the unidirectional virtual channel over which the current cells will be routed.

- *Payload Type Identifier (PTI)*: The three-bit PTI field is used to indicate network congestion and cell type, in addition to a number of

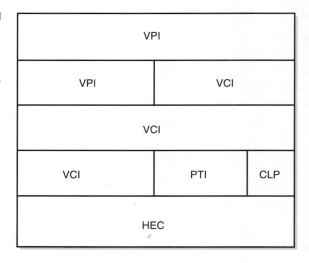

Figure 5-10
ATM cell header when used in NNI configuration.

other functions. Figure 5-11 illustrates these bits. The first bit indicates whether the cell was generated by the user or by the network, whereas the second indicates the presence or absence of congestion in user-generated cells or flow-related *operations, administration, and maintenance* (OA&M) information in cells generated by the network. The third bit is used for service-specific, higher-layer functions in the user-to-network direction, such as to indicate that a cell is the last in a series of cells. From the network to the user, the third bit is used with the second bit to indicate whether the OA&M information refers to segment or end-to-end-related information flow.

■ *Cell Loss Priority (CLP)*: The single-bit cell loss priority field is a relatively primitive flow control mechanism by which the user can indicate to the network which cells to discard in the event of a condition that demands that some cells be eliminated, and is similar to the *Discard Eligibility* (DE) bit in frame relay. It can also be set by the network to indicate to downstream switches that certain cells in the stream are eligible for discard should that become necessary.

■ *Header Error Control (HEC)*: The eight-bit HEC field can be used for two purposes. First, it provides for the calculation of an eight-bit *Cyclic Redundancy Check* (CRC) that checks the integrity of the entire header. Second, it can be used for cell delineation.

Figure 5-11
Payload Type
Identifier (PTI) field.

Value	Function
000	User data cell; no congestion; SDU=0
001	User data cell; no congestion; SDU=1
010	User data cell; congestion; SDU=0
011	User data cell; congestion; SDU=1
100	OAM F5 segment associated cell
101	OAM F5 end-to-end associated cell
110	Resource management cell
111	Reserved

ATM Services

The basic services that ATM provides are based on three general characteristics: the nature of the connection between the communicating stations (connection-oriented vs. connectionless); the timing relationship between the sender and the receiver; and the bit rate required to ensure proper levels of service quality. Based on those generic requirements, both the ITU-T and the ATM Forum have created service classes that address the varying requirements of the most common forms of transmitted data.

ITU-T Service Classes

The ITU-T assigns service classes based on three characteristics: connection mode, bit rate, and the end-to-end timing relationship between the end stations. They have created four distinct service classes based on the model shown in Figure 5-12. Class A service, for example, defines a connection-oriented, constant bit rate, timing-based service that is ideal for the stringent requirements of voice service. Class B, on the other hand, is ideal for such services as variable bit rate video, in that it defines a connection-oriented, variable bit rate, timing-based service. Class C service was defined for such things as frame relay, in that it provides a connection-oriented, variable bit rate, timing-independent service. Finally, Class D delivers a connectionless, variable bit rate, timing-independent service that is ideal for IP traffic as well as *Switched Multimegabit Data Service* (SMDS).

Figure 5-12
AAL classes of service.

	Class A	Class B	Class C	Class D
AAL Type	1	2	5, 3/4	5, 3/4
Connection Mode	Connection-oriented	Connection-oriented	Connection-oriented	Connectionless
Bit Rate	Constant	Variable	Variable	Variable
Timing relationship	Required	Required	Not Required	Not Required
Service Types	Voice, video	VBR voice, video	Frame relay	IP

In addition to service classes, the ITU-T has defined *ATM Adaptation Layer (*AAL) service types, which align closely with the A, B, C, and D service types described previously. Whereas the service classes (A, B, C, and D) describe the capabilities of the underlying network, the AAL types describe the cell format. They are AAL1, AAL2, AAL3/4, and AAL 5. However, only two of them have really survived in a big way.

AAL1 is defined for Class A service, which is a constant bit rate environment ideally suited for voice and voice-like applications. In AAL1 cells, the first octet of the payload serves as a payload header that contains cell sequence and synchronization information that is required to provision constant bit rate, fully-sequenced service. AAL1 provides circuit emulation service without dedicating a physical circuit, which explains the need for an end-to-end timing relationship between the transmitter and the receiver.

AAL5, on the other hand, is designed to provide both Class C and D services, and although it was originally proposed as a transport scheme for connection-oriented data services, it turns out to be more efficient than AAL3/4 and accommodates connectionless services quite well.

To guard against the possibility of errors, AAL5 has an eight-octet trailer appended to the user data which includes a variable size *pad field* used to align the payload on 48-octet boundaries, a two-octet *control field* that is currently unused, a two-octet *length field* that indicates the number of octets in the user data, and finally, a four-octet CRC that can check the integrity of the entire payload. AAL5 is often referred to as the *Simple and Easy Adaptation Layer* (SEAL), and it may find an ideal application for itself in the burgeoning Internet arena. Recent studies indicate that TCP/IP transmissions produce comparatively large numbers of small packets that tend to be around 48 octets long. That being the case, AAL5 could well transport the bulk of them in its user-data field. Furthermore, the maximum size of the user data field is 65,536 octets, which is coincidentally the same size as an IP packet.

ATM Forum Service Classes

The ATM Forum looks at service definitions slightly differently than the ITU-T. Instead of the A, B, –C, D, services, the ATM Forum categorizes them as real-time and non-real-time services. Under the real-time category, they define constant bit rate services that demand fixed resources with guaranteed availability. They also define real-time *variable bit rate* (VBR) service, which provides for statistical multiplexed, variable bandwidth service allocated on demand. A further subset of real-time VBR is peak-allocated VBR, which guarantees constant loss and delay characteristics for all cells in that flow.

Under the non-real-time service class, *Unspecified Bit Rate* (UBR) is the first service category. UBR is often compared to IP in that it is a best effort delivery scheme in which the network provides whatever bandwidth it has available, with no guarantees made. All recovery functions from lost cells are the responsibility of the end user devices.

UBR has two sub-categories of its own. The first, *Non-Real-Time VBR* (NRT-VBR), improves the impacts of cell loss and delay by adding a network resource reservation capability. *Available Bit Rate* (ABR), UBR's other sub-category, makes use of feedback information from the far end to manage loss and ensure fair access to and transport across the network.

Each of the five classes makes certain guarantees with regard to cell loss, cell delay, and available bandwidth. Furthermore, each of them takes into account descriptors that are characteristic of each service described. These include *peak cell rate* (PCR), *sustained cell rate (*SCR), *minimum cell rate* (MCR), *cell delay variation tolerance* (CDVT), and *burst tolerance* (BT).

ATM Forum Specified Services

The ATM Forum has identified a collection of services for which ATM is a suitable, perhaps even desirable, network technology. These include *cell relay service* (CRS), *circuit emulation service* (CES), *voice and telephony over ATM* (VTOA), *frame relay bearer service* (FRBS), LAN emulation (LANE), *multiprotocol* over ATM (MPOA), and a collection of others.

Cell relay service is the most basic of the ATM services. It delivers precisely what its name implies: a raw pipe transport mechanism for cell-based data. As such it does not provide any ATM bells and whistles, such as quality of service discrimination; nevertheless, it is the most commonly implemented ATM offering because of its lack of implementation complexity.

Circuit emulation service gives service providers the ability to offer a selection of bandwidth levels by varying both the number of cells transmitted per second and the number of bytes contained in each cell.

Voice and telephony over ATM is a service that has yet to be clearly defined. The capability to transport voice calls across an ATM network is a non-issue, given the availability of Class A service. What are not clearly defined, however, are corollary services such as 800/888 calls, 900 service, 911 call handling, enhanced services billing, SS7 signal interconnection, and so on. Until these issues are clearly resolved, ATM-based, feature-rich telephony will not become a mainstream service, but will instead be limited to simple voice—and there *is* a difference.

Frame relay bearer service refers to ATM's capability to interwork with frame relay. Conceptually, the service implies that an interface standard enables an ATM switch to exchange date with a frame relay switch, thus creating interoperability between frame- and cell-based services. Many manufacturers are taking a slightly different tack, however: they build switches with soft, chewy cell technology at the core and surround the core with hard, crunchy interface cards to suit the needs of the customer.

For example, an ATM switch might have ATM cards on one side to interface with other ATM devices in the network, but frame relay cards on the other side to enable it to communicate with other frame relay switches, as shown in Figure 5-13. Thus, a single piece of hardware can logically serve as both a cell and frame relay switch. This design is becoming more and more common because it helps to avoid a future rich with forklift upgrades.

LAN emulation enables an ATM network to move traffic transparently between two similar LANs, but also enables ATM to transparently slip into the LAN arena. For example, two Ethernet LANs could communicate across the fabric of an ATM network, as could two token-ring LANs. In effect, LANE allows ATM to provide a bridging function between similar LAN environments. In LANE implementations, the ATM network does not handle MAC functions such as collision detection, token passing, or beaconing; it merely provides the connectivity between the two communicating endpoints. The MAC frames are simply transported inside AAL5 cells.

One clear concern about LANE is that LANs are connectionless, whereas ATM is a virtual circuit-based, connection-oriented technology. LANs routinely broadcast messages to all stations, whereas ATM allows point-to-point or multipoint circuits only. Thus, ATM must look like a LAN if it is to behave like one. To make this happen, LANE uses a collection of specialized

Figure 5-13
Frame Relay Bearer
Service (FRBS).

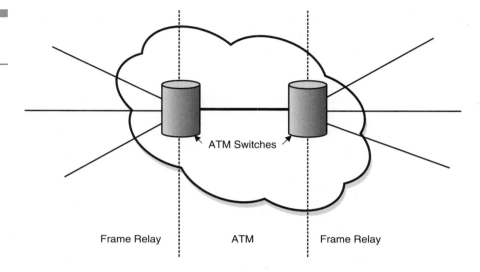

ATM Switches

Frame Relay   ATM   Frame Relay

LAN emulation clients and servers to provide the connectionless behavior expected from the ATM network.

On the other hand, multiprotocol over ATM provides the ATM equivalent of *routing* in LAN environments. In MPOA installations, routers are referred to as MPOA servers. When one station wants to transmit to another station, it queries its local MPOA server for the remote station's ATM address. The local server then queries its neighbor devices for information about the remote station's location. When a server finally responds, the originating station uses the information to establish a connection with the remote station, while the other servers cache the information for further use.

MPOA promises a great deal, but it is complex to implement and requires other ATM components such as the Private NNI capability to work properly. Furthermore, it's being challenged by at least one alternative technology, known as IP switching.

Originally developed by Ipsilon Corporation, now part of Nokia, IP switching is far less overhead-intensive than MPOA. Furthermore, it takes advantage of a known (but often ignored) reality in the LAN interconnection world: most routers today use IP as their core protocol, and the great majority of LANs are still Ethernet. This means that a great deal of simplification can be done by crafting networks to operate around these two technological bases. This is precisely what IP switching does. By using existing, low-overhead protocols, the IP switching software creates new ATM connections dynamically and quickly, updating switch tables on the fly. In IP switching environments, IP resides on top of ATM within a device, as shown in Figure 5-14, providing the best of both protocols. If two communicating devices wish to exchange information and they have done so before, an ATM mapping already exists and no layer three involvement (IP) is required—the ATM switch portion of the service simply creates the connection at high speed. If an address lookup is required, then the call is handed up to IP, which takes whatever steps are required to perform the lookup (a DNS request, for example). Once it has the information, it hands it down to ATM, which proceeds to set up the call. The next time the two need to communicate, ATM will be able to handle the connection.

Other services are looming on the horizon in which ATM plays a key role, including wireless ATM and video on demand for the delivery of interactive content such as videoconferencing and television. This leads to what I often refer to as "the great triumvirate:" ATM, SONET or SDH, and broadband services. By combining the powerful switching and multiplexing fabric of

Figure 5-14
IP switching.

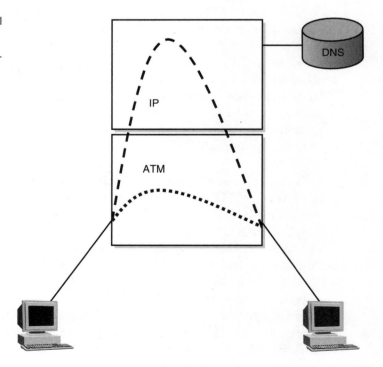

ATM with the limitless transport capabilities of SONET or SDH, true broadband services can be achieved, and the idea of creating a network that can be all things to all services can finally be realized.

ATM Protocols

Like all modern technologies, ATM has a well-developed protocol stack, shown in Figure 5-15, that clearly delineates the functional breakdown of the service. The stack consists of four layers: the Upper Services Layer, the ATM Adaptation Layer, the ATM Layer, and the Physical Layer.

The *Upper Services Layer* defines the nature of the actual services that ATM can provide. It identifies both constant and variable bit rate services. Voice is an example of a constant bit rate service, whereas signaling, IP, and frame relay are examples of both connectionless and connection-oriented variable bit rate services.

The *ATM Adaptation Layer* (AAL) has four general responsibilities:

- Synchronization and recovery from errors
- Error detection and correction

Figure 5-15
ATM protocol model (sometimes called the Broadband ISDN protocol model).

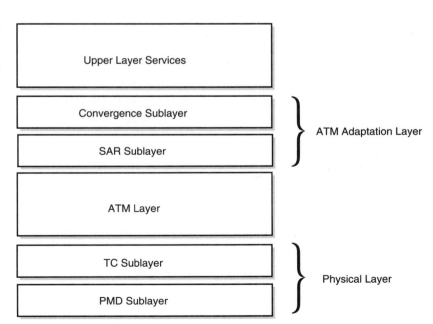

- Segmentation and reassembly of the data stream
- Multiplexing

The AAL comprises two functional sublayers. The *Convergence Sublayer* provides service-specific functions to the Services Layer so that the Services Layer can make the most efficient use of the underlying cell relay technology that ATM provides. Its functions include clock recovery for end-to-end timing management; a recovery mechanism for lost or out-of-order cells; and a time-stamp capability for time-sensitive traffic such as voice and video.

The *Segmentation and Reassembly Sublayer* (SAR) converts the user's data from its original incoming form into the 48-octet payload chunks that will become cells. For example, if the user's data is arriving in the form of 64-kilobyte IP packets, SAR chops them into 48-octet payload pieces. It also has the responsibility to detect lost or out-of-order cells that the Convergence Sublayer will recover from and to detect single bit errors in the payload chunks.

The *ATM Layer* has five general responsibilities. They are

- Cell multiplexing and demultiplexing
- Virtual path and virtual-channel switching
- Creation of the cell header
- Generic flow control
- Cell delineation

Because the ATM Layer creates the cell header, it is responsible for all of the functions that the header manages. The process, then, is fairly straightforward: the user's data passes from the Services Layer to the ATM Adaptation Layer, which segments the data stream into 48-octet pieces. The pieces are handed to the ATM Layer, which creates the header and attaches it to the payload unit, thus creating a cell. The cells are then handed down to the Physical Layer.

The *Physical Layer* consists of two functional sublayers as well: the Transmission Convergence Sublayer and the Physical Medium Sublayer. The Transmission Convergence Sublayer performs three primary functions. The first is called cell rate decoupling, which adapts the cell creation and transmission rate to the rate of the transmission facility by performing *cell stuffing*, which is similar to the bit stuffing process described earlier in the discussion of DS-3 frame creation. The second responsibility is cell delineation, which enables the receiver to delineate between one cell and the next. Finally, it generates the transmission frame in which the cells are to be carried.

The Physical Medium Sublayer takes care of issues that are specific to the medium being used for transmission, such as line codes, electrical and optical concerns, timing, and signaling.

The Physical Layer can use a wide variety of transport options, including

- DS1/DS2/DS3

- E1/E3

- 25.6 Mbps UNI over UTP-3

- 51 Mbps UNI over UTP-5 (*Transparent Asynchronous Transmitter / Receiver Interface* (TAXI))

- 100 Mbps UNI over UTP-5

- OC3/12/48c

Others, of course, will follow as transport technologies advance.

ATM Switch Design Considerations

The typical ATM switch consists of some form of high-speed switch fabric across which connections will be made, a processor that controls the goings-on within the switch, and a collection of input and output controllers that govern access to and from the network. Because it is a relatively slow function, the control processor is only used to establish connections, manage available bandwidth, and perform maintenance and management tasks. All cell processing is performed in the much faster input controllers, the switch fabric, and the output controllers.

ATM applications demand very high-speed switching, and the only way to achieve the speed required by these applications is to use hardware-based switching schemes in which software lookups play no role. All virtual path and channel translations are performed in silicon, as is the actual switching process.

ATM switches must also manage blocking processes, queuing performance, and routing; they must also be scalable, reliable, and testable in the event a failure occurs.

These considerations must be addressed during the design phase of any ATM switch. The blocking nature of the switch must be considered; the architecture and topology of the switch fabric itself must be carefully selected; and finally, the physical location of cell buffers must be carefully chosen because their location will determine the nature of the contention resolution scheme that is to be employed.

Two forms of switch fabric have emerged as acceptable solutions to the ATM speed dilemma. Time-division multiplexing is the simplest fabric to implement; in time-division systems, all cells are transmitted across a shared medium—usually either across a common bus or an array of shared memory—between the input and output ports. Both techniques work well, although the shared-memory systems are typically faster than those that employ the shared-bus architecture. The shared bus, however, is less expensive to implement.

The second form of switch fabric relies on space-division multiplexing, a technique in which large numbers of paths exist between input and output arrays. In what are called self-routing systems, the input controllers attach a routing tag to each cell using the same table lookup that they use to perform the VPI/VCI translations. Each element in the switch, then, uses the tag to route the cell accordingly.

In label-based routing, the VPI/VCI label is used to select specific routing tables within each element of the switch fabric. In theory, these work well and are non-blocking; however, they don't scale particularly well and are therefore not as efficient as self-routing switches.

Another issue that ATM switches must contend with is cell buffering. It is a well-known fact that although the provisioning of more buffers results in less cell loss and congestion, it also results in greater delay. Efficiency of use also plays a key role, so the method used to assign available buffers to cell streams is critical.

In internally buffered systems, memory is provided within the switch fabric so that the input and output queues share the available buffer pool. This results in a very efficient model, but although it is scalable, it has some difficulties providing cell prioritization and multicasting.

In externally-buffered systems, memory is physically located at either the input or output ports, or both. When input queuing is implemented, the buffer space is allocated at the ingress point, whereas with output queuing the buffers are placed on the egress side. Output queuing is generally considered to be more efficient and can be improved through the use of shared buffers, in which all output queues have access to a shared chunk of available memory. A feedback loop is provided for contention management purposes.

Other characteristics of ATM switches include management of switching delay and throughput, the number of user interfaces and ports available, AAL and QoS support, support for both permanent and demand virtual channels, point-to-multipoint support, congestion control across the network, and a number of other options.

All in all, the technology behind ATM switch architectures is reasonably mature and improves routinely. The number of players in the manufacturing sector continues to grow, and as both the market requirements and the underlying technology are better understood, the switch business has become quite competitive, leading to a variety of reliable, innovative, and capable products to choose from.

The Players

The players on the ATM field include the edge and backbone switch manufacturers, as well as the service providers themselves. The ATM switch market has segmented itself into two product areas: the edge switches, which are analogous to local central office switches in the telephony domain, and core switches, which are the equivalent of tandem switches. Edge switches, as their name implies, are located at the edge of the network, and as such provide the access point for users entering the network. They fall into two categories. True ATM edge switches permit the connection of non-ATM devices to the network, whereas ATM access switches are designed to enable other ATM devices to connect. Among others, Lucent, Nortel, and 3Com all manufacture edge switches, with reasonably similar features. All offer support for a wide variety of media and LAN schemes, as well as ATM access ports and support for multiple virtual LANs. The price per port varies somewhat, but not unreasonably so.

Furthermore, these switches support a wide (and growing) variety of ATM features, such as IP over ATM, LANE, ATM UNI, and the full complement of service levels such as ABR, CBR, UBR, and so on. In the LAN environment, these switches support traditional Ethernet, Fast Ethernet, Gigabit Ethernet, and typically rely on bus architectures within the switch.

Core or backbone switches, on the other hand, which are the tandem switches of the ATM realm, are quite a bit more robust than the edge switches, with throughput as high as 13 Gbps. The price-per-port is comparable to that of the edge switches, but the feature and services complement that they offer is somewhat expanded. In addition to the features supported by the edge switches, ATM core switches also support LAN emulation client and server, flow priority control, and extensive network management capabilities. Leading switch manufacturers in the marketplace include Lucent, Cisco, and IBM, among others.

ATM Service Offerings

Plethora is a great word for describing the diversity of players in the ATM services provisioning game. The traditional service providers (ILECs, CLECs, and IXCs) support a wide variety of physical interfaces including T1, DS3, OC3, OC-12, OC-48, and OC-192. Most support both CBR and VBR, frame-relay interworking, and port oversubscription.

ATM is a technology that continues to be deployed in a broad range of industries because it works well. One could say that ATM is like a well-behaved child: "it plays well with others." It is far from being the ideal technology solution for the transport of any particular protocol or service; after all, the *Public Switched Telephone Network* (PSTN) is the best solution for voice, whereas a dedicated network is better for video. For the broad mix of multiple protocol types, however, ATM is ideal. In that sense, it is a lot like a duck. A duck walks, swims, flies, and makes bird noises, but does none of them particularly well—but it does do them all.

In Summary

So let me summarize the main points of this chapter. First of all, ATM remains a robust, capable technology, with a great deal of promise in the evolving transport and switching realm. It appears that its role will evolve to that of an edge technology, sitting at the periphery of the network cloud, serving as an aggregation technology for diverse traffic types for transport across what will most likely be an IP-based wide area network fabric. The ATM standards and services are largely mature, and offer granular control over quality of service measures, class of service assignment, and bandwidth management. Because of its cability to aggregate and therefore simplify network architectures, it will help service providers avoid the Winchester Mystery House networking problem, while still offering a diversity of service types across a single, technologically simple, robust cloud. All the players are here—service providers, manufacturers, and customers are working well together to develop and support applications that run well across ATM and that take advantage of its unique capabilities. Furthermore, it continues to evolve as a key technology in the ever-expanding realm of "everything over IP."

In short? Keep watching: the game ain't over!

Players and Futures in the SONET/SDH Game

For the past 13 years, I have spent the majority of my professional life studying, writing, and teaching about the telecommunications industry and the companies that comprise it. It has always been a matter of some fascination to me how those players' roles have evolved over time as they adapt to the changing winds of the telecommunications marketplace and the chaotic evolutionary track of the various technologies that periodically evolve from those that came before.

SONET and SDH live in a segment of the telecommunications industry that is devoted to one thing and one thing only: improving the financial success of the companies that build SONET and SDH networks. These include *incumbent local exchange carriers* (ILECs) and PTTs, *competitive local exchange carriers* (CLECs) and city carriers, long-distance carriers, and more and more, the players in the non-traditional transport industry including cable television, power, pipeline, railroad, and anyone else with underutilized or redeployable right-of-way. Think about it: Sprint and Qwest both emerged as spin-offs from the railroad industry; WorldCom has its roots in Oklahoma's natural gas pipelines; and any number of transport providers have emerged from power companies all over the world.

We have already seen that bandwidth demand is at an all-time high and will continue for the foreseeable future. We also know that the generic transport services that the incumbents, new entrants, and long-distance companies provide are rapidly becoming commodities, available from any number of providers at rock-bottom prices. In classical supply and demand terms, as supply goes up, demand goes down—and in another model, as supply goes up and commoditization takes place, the price plummets, as shown in Figure 6-1. When the price of a commodity drops precipitously, margins on that product disappear. This is precisely the conundrum that bandwidth providers face. Their principal deliverable, bits per second, is offering rapidly accelerating and diminishing rates of return, which portends one thing: if those companies wish to remain in business and solvent, they must offer enhanced services to their customers above and beyond simple bandwidth. That's the increase revenue part of the equation described previously. Furthermore, they must come up with a way to offer that bandwidth in such a way that even in the face of precipitous margin decline, it remains a profitable offering. That's the reduce costs part of the equation.

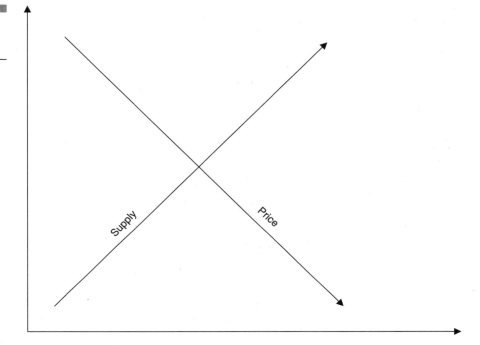

SONET and SDH, as it turns out, are squarely positioned to address both of these issues. Consider the evolution of the typical optical network. When service providers first began to build long-haul optical transport systems in the late 1970s and early 1980s, the networks were largely point-to-point, as shown in Figure 6-2, because SONET, SDH, and the concept of a ring architecture had not yet arrived. Traffic that needed to be transported between Dallas and Denver, for example, would travel across a dedicated point-to-point optical link between the two end points. It wasn't the least bit flexible, but it offered bandwidth at high volumes and satisfied the nascent demand for something beyond copper transport. The problem with this dedicated model was twofold. First, it was highly vulnerable. A carefully situated backhoe digging in the wrong place could easily disrupt massive volumes of sensitive corporate traffic as well as voice, video, and other customer information streams. Recovery meant either a lengthy dispatch and repair job or some form of temporary reroute via digital or manual cross-connect.

Figure 6-2
A point-to-point circuit.

The second problem was that the model was limited in capacity. Naturally, the optical span offered tremendous bandwidth, but everything is finite, including an optical fiber. If the span were to be exhausted, the only solution was to provision additional spans, illustrated in Figure 6-3. This solution had two key downsides. First, it was inordinately expensive. To add fiber capacity, the trench had to be opened, the fiber, amplifiers, and repeater equipment had to be installed, and the hole had to be filled again. That process involved enormous cost and time because of the human-intensive nature of the installation process. A solution came in two forms, one well before the other. The first was optical *time-division multiplexing* (TDM), handled well by SONET and SDH, which enabled service providers to transport multiple data streams across a facility by assigning time slots to them. The second was *Dense Wavelength Division Multiplexing* (DWDM), which permitted service providers to transport multiple higher bit rate streams by assigning wavelengths to them, facilitating extremely high bit rate transport. SONET/SDH and DWDM, then, played a major role in early cost reduction efforts.

Network Evolution Begins

The next stage in the evolutionary process saw linear networks give way to rings as the criticality of high-speed network transport became obvious and the capabilities of the SONET/SDH overhead are realized. You will recall that the *K* bytes give SONET and SDH the capability to monitor the health of the network connections between multiplexers, the multiplexers themselves, and in the event of a failure to provide automatic protection switching as a way of surviving a cable cut. As the SONET/SDH overhead became better understood, service providers began to explore alternative network architectures that could take best advantage of the capabilities of that overhead. Two architectures in particular emerged: the two- and four-fiber ring.

Figure 6-3
Multiple circuits.

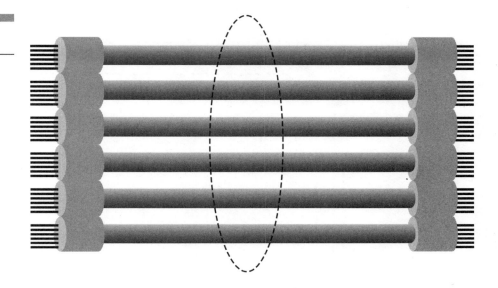

Ring Architectures

As we described earlier, ring architectures have several advantages. The first is the redundancy and survivability that the architecture provides. In a two-fiber implementation, shown in Figure 6-4, the rings are typically designated as a primary or active ring and a protect or backup ring. Live traffic is carried on the active path under normal operations, whereas the backup ring simply monitors the status of the overall system and transports keep-alive messages. In the event of a failure of the primary ring, automatic protection switching kicks in, causing the devices on either end of the ring breach to switch traffic (usually within 50 ms) to the backup span. Under normal circumstances, the user will be unaware of the switchover. This particular two-fiber architecture is known as a *Unidirectional Path-Switched Ring* (UPSR).

In the event of a dual-span failure (due to the efforts of our overly diligent backhoe driver described in Chapter 2), the ring wraps at the multiplexer nodes on either side of the breach, resulting in the creation of a single span ring, but preserving the integrity of the transmission path. This is shown in Figure 6-5.

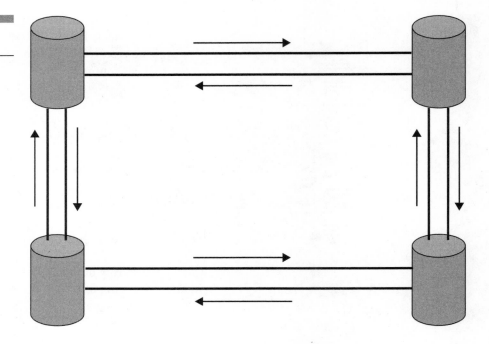

Figure 6-4
A two-fiber ring.

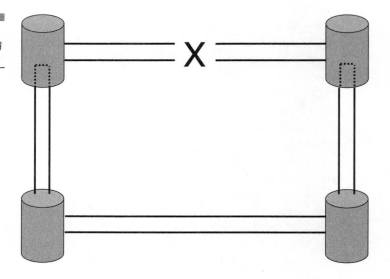

Figure 6-5
Ring wraps following
breached fiber.

Two-fiber rings can also be configured in such a way that active traffic travels on both rings simultaneously. This is known as a *Bidirectional Line Switched Ring* (BLSR). Let's assume that our ring is carrying 16 SONET STS or SDH STM channels. In a BLSR configuration, eight of the channels would travel on the clockwise ring, while eight would travel in the opposite direction on the counter-clockwise ring. Under normal operating circumstances, any single payload component would simply travel from the originating multiplexer to the terminating multiplexer, as shown in Figure 6-6.

Should a failure of both rings occur between two multiplexers, the ring would wrap, as shown in Figure 6-7. Note that the result is the same: the traffic will take longer to reach its destination (which could be a problem for some applications), but it will reach the destination nonetheless. Be aware that the only devices that are aware of the nature of the failure in this situation are the two multiplexers on either side of the fiber cut that have to respond to the failure.

Four-fiber rings, on the other hand, rely on a slightly different architecture, shown in Figure 6-8. A four-fiber ring offers greater, more varied flexibility. For example, let us assume that fibers A and B are serving as the active path, while fibers C and D are serving as the backup. Because of added capacity, we can now transport 16 channels on the clockwise path (A) and 16 more in the counter-clockwise direction (B), as shown in Figure 6-9.

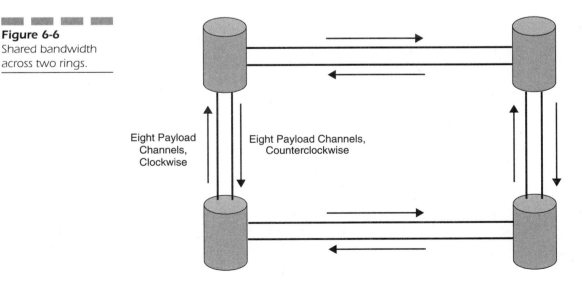

Figure 6-6
Shared bandwidth across two rings.

Eight Payload Channels, Clockwise

Eight Payload Channels, Counterclockwise

Figure 6-7
Ring recovery.

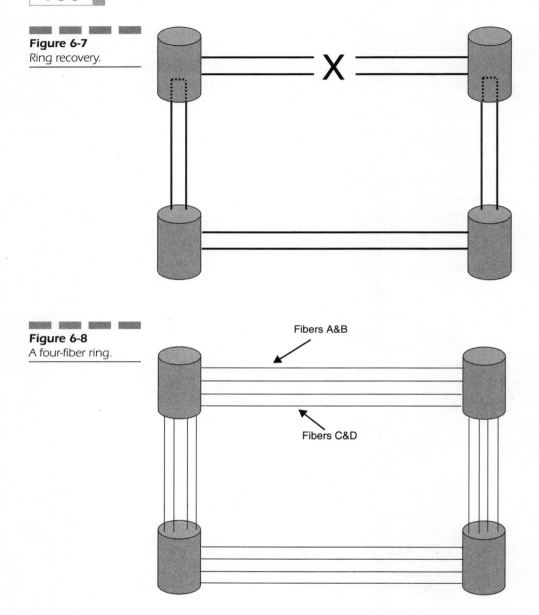

Figure 6-8
A four-fiber ring.

Fibers A&B

Fibers C&D

Paths C and D remain quiescent. In this case, a payload component would travel from the originating multiplexer to the terminating multiplexer in a relatively linear fashion around the ring, as shown in Figure 6-10.

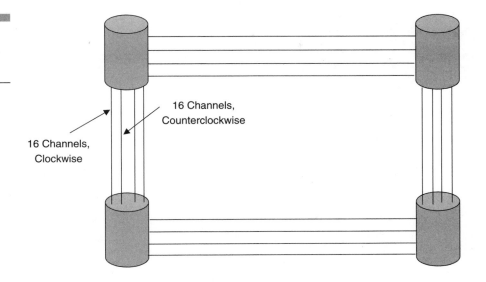

Figure 6-9
Shared bandwidth across a four-fiber ring.

16 Channels, Counterclockwise

16 Channels, Clockwise

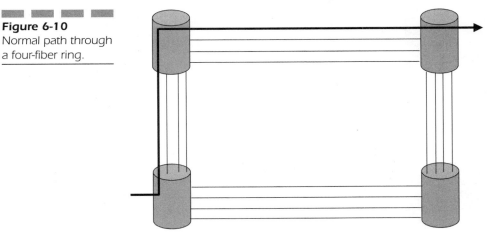

Figure 6-10
Normal path through a four-fiber ring.

Should the two primary fibers fail, as shown in Figure 6-11, the ring recovers in exactly the same way the two-fiber ring did by routing traffic to the backup span, but only to avoid the failed area. Again, only the devices on either end of the failure are aware that the failure has occurred.

Should all four fibers fail, as illustrated in Figure 6-12, the BLSR has the capability to recover. In this case, the active and protect spans work

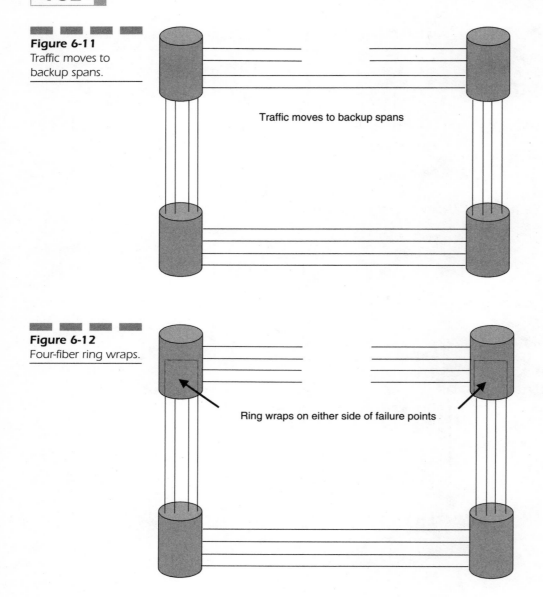

Figure 6-11
Traffic moves to
backup spans.

Traffic moves to backup spans

Figure 6-12
Four-fiber ring wraps.

Ring wraps on either side of failure points

together to route around the failure point, converting the ring into a toroidal shape and guaranteeing survivability and traffic delivery.

Ring architectures, then, provided a significant improvement over dedicated, point-to-point designs in terms of flexibility, survivability, customer service, and revenue protection. Furthermore, as optical technology advanced, the bandwidth available over a ring grew from 2.5 Gbps to 10 Gbps to 40 Gbps (and potentially beyond). Furthermore, the addition of DWDM to an otherwise pure SONET or SDH ring added yet another layer of transport capability that dramatically multiplied the overall capacity of the architecture. This became particularly valuable with the increasing focus on metropolitan applications.

The Metro Explosion

The metro region has enjoyed a great deal of technological attention in the last 18 months because service providers have realized that this region of the network has an enormous unmet need, and the service providers, along with their manufacturer partners, have the wherewithal to address it. The metro business market relies heavily on LAN technology, largely the various flavors of Ethernet (traditional 10-Mbps, 100-Mbps Fast Ethernet, and 1,000-Mbps Gigabit Ethernet) for the transport of corporate data among workers in one or more locations. The problem they face today is the result of changing corporate work models. More and more, the concept of the single, monolithic corporate headquarters building is going away in favor of numerous smaller work locations that are located closer to the customer and closer to suburban areas where employees live, illustrated in Figure 6-13. The need to share information among the people in those locations has not changed (in fact, driven by the awareness of the strategic value of shared information and interest in knowledge management, it may actually be greater), so cost-effective solutions are needed to provide a way to transport Ethernet traffic across the metropolitan area. SONET and SDH have emerged as clear contenders for this responsibility, and manufacturers have announced Ethernet-to-SONET/SDH-to-Ethernet products that meet the demand.

From Rings to Meshes

Rings are wonderfully innovative network solutions, but they do have limitations. First, in spite of all the capabilities they bring to the network, they are not particularly flexible in terms of deployment. Users of them are forced to transport their traffic along the single path that the ring provides,

Figure 6-13
The metro
environment.

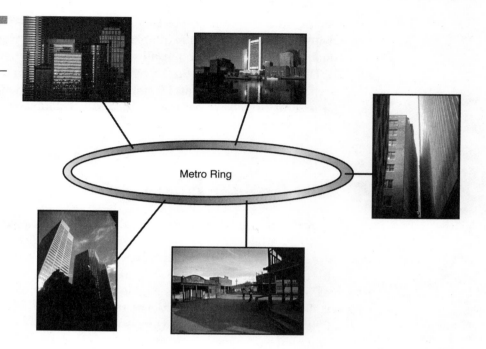

even if that path is a long way around from the source to the destination of the transmission. In many cases, this poses no problem, but for the many delay-sensitive applications that are emerging on the scene, such as voice over IP, interactive and distributive video, and others, the added delay can change the perceived quality of service, especially if a ring switch has occurred, that adds even more end-to-end delay. In response to this evolving perception, a new architecture has emerged on the scene. Interestingly enough, it is not a new model, but rather one that has been around for a long time. *Mesh architectures* are suddenly enjoying great attention as the next generation of networking to inherit the crown from the pure ring. This model, shown in Figure 6-14, offers the multiple route capabilities of a dedicated network span alongside the survivable nature of a ring. The bulk of the network is clearly a collection of point-to-point facilities that provide shortest distance transport between any two end points, as illustrated by Figure 6-15. However, the collection of dotted lines clearly represents a ring, and the large number of alternate routes means that survivability is assured alongside guarantees of least-hop routing. In the real world, fiber

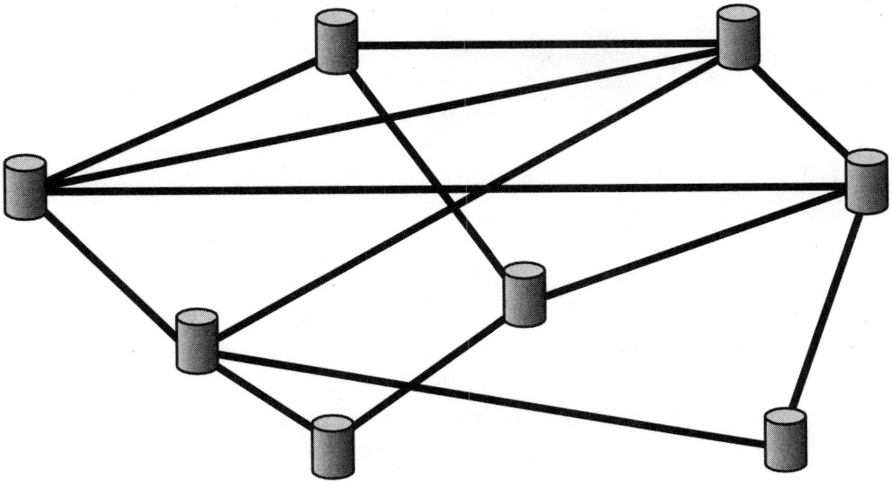

Figure 6-14
A mesh network.

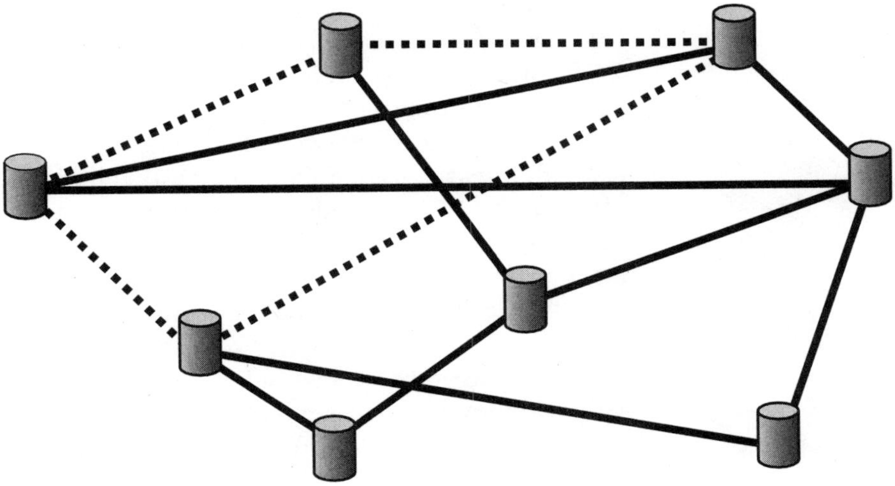

Figure 6-15
Logical rings in a
mesh environment.

has now been deployed so extensively and universally that the ability to build mesh networks has not become a problem.

The real issue with SONET and SDH as they evolve to accommodate the needs of the so-called next-generation network is this: both must be modified in some way to transport the variety of data types that have emerged as viable revenue components in the emerging optical network. SONET and SDH were designed to address the relatively predictable 64-Kbps transport needs of voice networks that dominated the attention of service providers in the pre-Internet world. However, with the arrival of the World Wide Web in the early 1990s, transported traffic became a mix of predictable voice and unpredictable, chaotic, latency-friendly packet data. As a result, SONET and SDH were no longer capable of meeting the demands of all service types as adequately as customers required, particularly given the nascent interest in *quality of service* (QoS) as the principal differentiator among access and transport providers. Qualities such as security, latency, granular bandwidth provisioning, dynamic time-of-day provisioning, multiple levels of service protection, and a host of others have garnered the attention of service providers in general, particularly as they have begun to managerially segment their networks into local, metro, regional, and long-haul quadrants. It is this differentiable quality of service capability that not only provides differentiation among the players in a rapidly commoditizing (if that's a word) market, but it also offers new approaches to revenue generation—always a happy topic.

Because of the original services that they were designed to transport, SONET and SDH networks for the most part comprise large, multinode rings that interconnect to other rings as well as point-to-point facilities. These architectures are well understood, fully functional, and widely deployed, thus the cost of maintaining them on an ongoing basis is comparatively low. They provide ideal carriage for the limited requirements of circuit-switched voice, offering not only low-latency transport, but survivability as well. If a failure occurs due to a fiber disruption, service can usually be restored in less than 50 ms, which means that voice traffic is not affected.

As the traffic mix has evolved, however, the limitations of SONET and SDH have become rather more evident. Each ring in a SONET or SDH network is limited to a certain number of nodes, and each ring cannot exceed a certain maximum circumference if transport QoS is to be assured. Thus, if added capacity is needed in the network, the solution is to add rings, clearly an expensive and time-consuming process. This stacked ring model also has the disadvantage of being disparately managed, meaning that each ring must be provisioned and managed on an individual basis, making the job of the network manager rather more difficult.

Furthermore, SONET and SDH do not offer particularly flexible bandwidth allocation capabilities. Originally created to transport 64-Kbps voice, neither of the two adapts well to the transport of data traffic that has wildly variable bandwidth requirements, in terms of both bandwidth assignment and predictability. Think about it: in a SONET network, bandwidth jumps from 51.84 Mbps (OC-1) directly to 155.52 Mbps (OC-3) with no intermediate level. A customer wishing to make a minor upgrade must increase his or her actual purchased bandwidth threefold, much of which he or she will probably not use. And because SONET and SDH rings typically reserve as much as half of their total available bandwidth for redundancy functions, they are *terribly* inefficient.

The mesh model introduced earlier represents the solution to this set of problems. In the last 3 years, optical networking has evolved in three significant areas: the development of true all-optical switching and intelligent routing, the extensive proliferation of fiber throughout most carriers' operating areas, and the return of the mesh network.

In a ring network, nodes are connected to one another in such a way that they do not have direct interconnection to one another—all node-to-node traffic (other than between adjacent nodes) must flow along a rigidly deterministic path from a source node to a destination. In a mesh network, every node in the network is connected to every other node in the network, thus enabling shortest-hop routing throughout the network between any two end points.

The advantages of this design are rather strong. First of all, distance limitations in mesh networks are largely eliminated because paths are created on a shortest-path basis between nodes rather than all the way around a ring. As a result, nodes represent the bottleneck in mesh networks rather than the fiber spans themselves, which means that network operators can increase capacity simply by adding nodes on a demand basis and increasing transported bandwidth across the installed fiber infrastructure. This eliminates the stacked ring problem and dramatically improves upgrade intervals, an area of some concern for most service providers.

Perhaps the greatest advantage of the mesh deployment is management: unlike SONET and SDH, where rings must be managed largely on a ring-by-ring basis, mesh networks are designed to accommodate point-and-click provisioning, which shortens installation intervals from months in many cases to hours. A number of vendors, including Astral Point, have put themselves on the map with this capability. Their focus is primarily on the metropolitan marketplace, whereas players such as Sycamore have focused their attention more on the wide area transport environment.

The mesh network also enables carriers to design *distributed protection* schemes, which in turn enable them to use their available bandwidth much more efficiently. SONET and SDH rely on redundant rings to provide 100 percent survivability, which on the one hand is great because it provides 100 percent redundancy, but on the other hand wastes 50 percent of the available bandwidth because of the reservation of an entire ring for the eventuality of a rogue backhoe driver with a map of network routes in his or her back pocket. Because of the tremendous flexibility of mesh topology, entire routes do not need to be reserved. Instead, network designers can allocate backup capacity across a collection of routes, thus preserving the integrity of the network, while at the same time using the available bandwidth far more efficiently than a ring. Furthermore, most carriers have plans to offer a variety of protection levels to their customers as one of many differentiable services; SONET and SDH, with their 100 percent, always-on protection schemes do not lend themselves to this capability as a mesh architecture, whereas a mesh can be provisioned in any of a variety of ways. These include 0:1 protection (no protection at all), 1:N protection (one protect path that handles the potential failures of multiple paths), and 1:1 protection (dedicated protection a la SONET and SDH rings). The first is obviously the most economical, whereas 1:1 protection is the most costly because it offers 100 percent protection against network failures. Obviously, mesh networks are far more complex than their ring counterparts and are therefore significantly more difficult to manage. They require a level of network management intelligence that is quite a bit more capable than that provided by the SONET and SDH overhead bytes.

Both ring and mesh networks have their advantages and disadvantages, as we have just discussed. Furthermore, SONET and SDH enjoy an enormous embedded base of installed networks and service providers that have made large capital investments to build them are loath to discard those investments easily. Not only must hardware be modified, but the evolution from ring to mesh involves modifications to provisioning systems, billing systems, operations support, provisioning, and personnel training as well. The result of this is that most industry analysts believe that the end result, at least in the near to medium term, is the evolution of hybrid networks that sport the best of both ring and mesh architectures, shown in Figure 6-16. Nevertheless, the demand for both will continue.

So when does one offer more than the other? When does a SONET or SDH ring outperform a SONET or SDH mesh? The answers are varied, but a few guidelines emerge repeatedly in analyses of the two. These are described in the following section.

Fiber Issues

First of all, because mesh networks are more efficient allocators of bandwidth than rings, they represent a better choice in carrier networks where fiber exhaust or potential fiber exhaust may become an issue.

Figure 6-16
A hybrid mesh/ring
architecture.

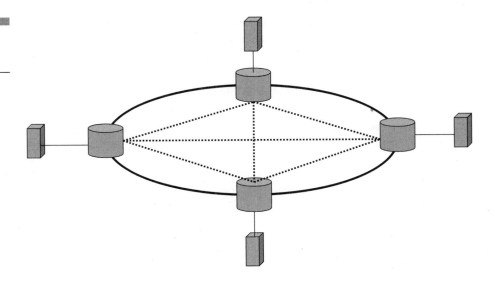

Network Geography

In the core of the network, where long-distance, often less differentiated traffic is transported, mesh networks will continue to be deployed. The farther a network technology is installed from the actual customer and his or her ever-evolving applications, the less differentiated it needs to be. Thus, the network model that serves core transport is more concerned with volume than it is with differentiable service types. Mesh, then, represents an ideal transport model for the network core.

Rings, on the other hand, will find a home along the margins of the network and within the metro arena. Bandwidth in the metro region tends to be generally cheaper than that deployed in the core. Furthermore, the need to reserve bandwidth for restoration guarantees is less of an issue in the metro area (and where mesh layouts are less common) because bandwidth is cheaper. In the long-haul domain, however, where bandwidth is far more expensive, mesh architectures provide a more efficient way to use bandwidth while simultaneously offering high-volume transport and guaranteed service.

The typical metro network environment comprises relatively low-speed access rings (155.52 Mbps to 2.488 Gbps) that aggregate and transport traffic from digital loop carrier equipment, corporate PBXs, and other customer devices that generate and terminate traffic. In the central office, the traffic is passed through a digital cross-connect system of one kind or another that grooms the traffic before passing it along to higher-speed aggregation rings. These aggregation rings interconnect to carrier *points-of-presence* (POPs), often in a metropolitan area, which in turn allow for hand off to wide area network connections. This is illustrated in Figure 6-17.

Traffic Characteristics

The type of traffic being transported across a network can be a deterministic factor where network designers are considering deployment of ring vs. mesh architectures. In small geographic areas, rings are well suited to the transport of delay-sensitive traffic. Metro areas do not typically enjoy mesh penetration, and because most customers are connected to nodes on access rings rather than to nodes on a mesh, the model holds. In access applications, SONET and SDH rings rise to the top of the capability mountain; mesh, shown in Figure 6-18, shines when used in interoffice implementations.

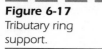

Figure 6-17
Tributary ring
support.

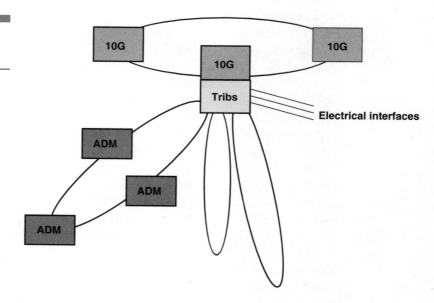

For wide area transport, mesh networks provide ideal service characteristics, particularly for the requirements of latency/delay-friendly data traffic where volume may be very high, but potentially bursty as well.

Defining the Network Edge

The edge of the optical network has multiple definitions, but it can be somewhat defined as the interface between the core and the customer. The type

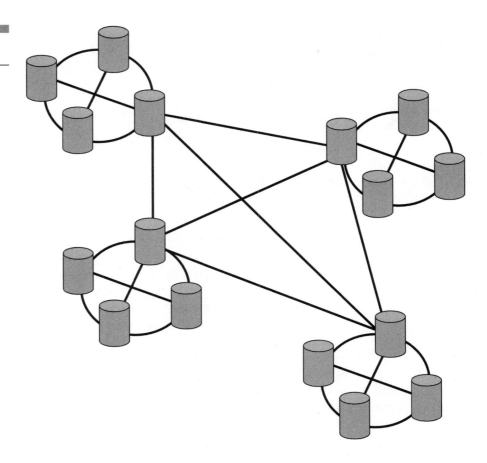

Figure 6-18
A mesh network.

of service provider also has some bearing on the definition of this particular network region because their responsibilities vary depending where they sit in the pantheon of network operators. In the following section, we will examine the major operators and their views of the network. First, however, we should describe the nature of the device that performs the optical edge function. It is typically a device that performs *add-drop multiplexing* (ADM) capability, multichannel transport through the use of DWDM, and some form of switching and/or routing. It must offer capacity that is *at least* equivalent to the bandwidth in a single DWDM lambda, typically 2.488 Gbps. This may provide a sticking point for some customers: ATM is one of the most widely transported protocols over SONET and SDH networks. ATM chipset manufacturers, however, typically build their devices to operate no faster than 622.08 Mbps. This clear disconnect could pose serious problems for operators looking to interface the two efficiently. It is, however, being addressed as awareness is very high.

Let's consider now the companies in the telecommunications industry that are most affected by this evolving technology model. They include the *Incumbent Local Exchange Carriers* (ILECs), the *Competitive Local Exchange Carriers* (CLECs), the *Interexchange Carriers* (IXCs), the Bandwidth Barons, the *Internet Service Providers* (ISPs), and a few others. All stand to benefit from the ongoing optical evolution; however, because they play in wildly differing markets, their issues are also different.

The Incumbent Local Exchange Carrier (ILEC)

The typical ILEC (formerly RBOC) operates a network that is hierarchical, comprising multiple levels of central offices that are interconnected by optical facilities that in turn feed bandwidth to *multitenant units* (MTUs) or individual customers over fiber or copper distribution facilities. In their minds, the edge of the network is either the central office or the end terminal located at the customer premises.

ILECs tend to be relatively homogeneous in terms of the products and services they provide. Their strengths lie in the access and transport business, at which they excel; however, because of the commoditization of this business, their ability to maintain marketshare is diminishing. For the most part, they have grown by acquiring more of the same: Witness Bell Atlantic's acquisitions of NYNEX and GTE, or SBC's acquisitions of Pacific Bell, Nevada Bell, Ameritech, and SNET. They have expanded their footprint, but have not done much to diversify their product and service offerings. In fairness, they have good reasons for this strategy; the ILECs have realized that with long-distance relief pending, they must create a wide area presence for themselves. Their larger business customers are not necessarily local companies; although they may have a local presence, they tend to be national or even global. If the ILECs are to become full-service *service providers*, they must be able to serve those customers on an end-to-end basis, thus eliminating the need for intermediaries. Without a wide area data network, they cannot accomplish this. The fact is the market is the ILECs' to lose. Many analysts believe that customers will buy all services from the local service provider, if the local service provider has the ability to provision them. The holder of the access lines rules; consequently, much of the company convergence activity of late has revolved around acquisition of access lines. Consider Qwest's acquisition of USWest or Global Crossing's acquisition of Frontier. On a slightly different level, AT&T's acquisition of TCI is clearly a gambit for local loops and more will follow.

So are ILECs a dying breed? Will they be brought down by the smaller, more nimble CLECs that are nibbling away at their longstanding customer bases? There is no question that they face some serious challenges. Their networks were designed around the idea that they would control 100 percent of the market and are therefore not the most cost-effective resource in an open and competitive market. Other models are far more cost-effective than the ILECs' circuit-switched infrastructures. As a result, the ILECs are reinventing themselves a piece at a time and are, of course, expanding their market presence in a variety of ways.

ILECs stand to benefit from the evolving nature of the network edge primarily due to simplification. By reducing the number of layers and network elements that they manage, they will be able to reduce provisioning times and improve overall customer service quality. As their customers clamor more for IP services and less for legacy circuit-switched transport, the evolution will enable them to migrate efficiently to a common network fabric for all services in response to the call for convergence. Furthermore, they will enjoy enhanced efficiency as they move away from the largely overprovisioned networks that are characteristic of SONET and SDH.

The Competitive Local Exchange Carrier (CLEC)

Because of their overall focus on the metro area, CLECs will modify their networks to enable them to efficiently interconnect with interexchange carrier *points-of-presence* (POPs) so that they can offer a full suite of services to their customers. In their minds, the edge of the network is slightly different than their ILEC relatives—they see it as the ILEC central office where they interconnect, the collocation cage, or the customers' equipment closet where they terminate their service.

The CLECs are similar to the ILECs in that they sell a commodity. They differ from the ILECs, however, because they tend to sell the more lucrative products and avoid the markets and services that don't enjoy high returns. For example, many CLECs focus on residential voice customers, whereas others go after business customers. All CLECs are not created equal; their business strategies and business plans for carrying out those strategies and satisfying customers vary dramatically from company to company. They are, however, good performers within their identified market niches. The CLECs are obviously after the same access lines that the ILECs want to protect; the ongoing convergence of medium-size CLECs is nothing more than a positioning move.

CLECs face significant obstacles by virtue of the fact that they are CLECs. As alternatives to the ILECs, they rely on interconnection agreements with them because they must have a collocation presence within the ILECs' central offices to provide service. The 1996 Telecommunications Reform Act mandates that before the ILECs will be allowed into the long-distance market, they must demonstrate that they have opened their local market to competitors, providing equal access to unbundled facilities, such as local loops and certain services. CLECs often complain that, although the ILECs have agreed to the stipulations, they are not particularly quick to respond to CLEC requests for interconnection service and therefore have the ability to exert some control on the pace at which CLECs can enter their markets. Of course, some checks and balances are in place, such as Section 251 of the 1996 Communications Act. This component of the law requires that ILECs sell circuits, facilities, and services to their competitors that are "at least equal in quality to that provided by the local exchange carrier to itself or to any subsidiary, affiliate, or any other party to which the carrier provides interconnection."

The ILECs control 90 percent of the access lines in the United States, so CLECs face a significant challenge. Many CLECs claim to be able to offer better, more customized service than their ILEC competitors. Most customers agree that the technology products sold by the ILECs and the CLECs are identical. The difference, they claim, is the way they deal with their customers. CLECs believe themselves to be more customer focused, claiming that the ILECs are still plagued by legacy monopoly mentality.

Whatever the case, some CLECs have initiated differentiation programs to help them garner the favor of customers, such as payback plans for downtime, online real-time usage reports, and negotiated service level agreements.

Ultimately, the success of the CLECs relies on three critical success factors. Local number portability must be viable, functional, and available; operations support standards must be in place and accepted; and discounts for unbundled network elements from ILECs must be on the order of 50 percent.

Like the ILECs, the CLECs stand to benefit greatly from the creation of the optical edge. They too will benefit from a simplified network architecture that will in turn lead to better, faster customer service. It will also enable the efficient migration to a packet network that will satisfy the demands of IP customers.

The Interexchange Carriers (IXCs)

Today's IXC has a multitiered network with central offices interconnected by DWDM-enhanced optical spans. They often employ ATM as their switching protocol and as a way of provisioning granular service quality levels. In their eyes, the edge of the network is the metropolitan hub point where multiple disparate services are aggregated for transport across high-speed backbone facilities.

The interexchange carriers face the greatest challenge of all the players, but are also the companies demonstrating the most innovative behavior in the face of adversity. With companies like Qwest and Level 3 building massively overcapacitized fiber networks, bandwidth is becoming so inexpensive and so universally available that it is evolving to a true commodity. The margins on it therefore are dropping rapidly. Furthermore, the number of companies that have entered the long-distance market has grown, as has their diversity: although the legacy players (AT&T, Sprint, and WorldCom) continue to hold the bulk of the market, a collection of power companies, satellite providers, and bandwidth barons have entered the game and are seizing significant pieces of marketshare from the incumbents.

In response, the IXCs are fighting back by diversifying. All have entered the ISP game, offering Internet access across their backbones at competitive prices. They have also bought or built local twisted pair access infrastructures, cable companies, satellite companies, and a host of others.

Consider AT&T as an example. Beginning with their long lines division, the company has grown into a multifaceted powerhouse that now owns a broad collection of companies offering diverse network services. They are a long-distance, local, wireless, ISP, portal, cable company, and have aggressive plans to be a true, full-service telecommunications provider as they flesh out their strategy for market positioning.

The optical edge provides a benefit to the IXCs in a variety of ways. It enables them to reduce their managed protocol layers (SONET/SDH, ATM) as a way to improve service provisioning and ongoing management, and helps them improve their cost effectiveness by reducing the total number of managed elements they must deal with.

The Bandwidth Barons (aka Carriers' Carriers)

These companies make their money by efficiently provisioning and selling bandwidth as a commodity to large volume customers. Their network is typically a DWDM backbone, often running ATM at Layer 2 and IP at Layer 3, that interconnects major metropolitan hub locations where traffic can be aggregated at enormous volumes for transport across the backbone. For these companies, the edge of the network is typically at these aggregation points where individual DWDM lambdas are connected to add-drop multiplexers and core network routers, sometimes running MPLS.

These companies used their access to right-of-way to build massive, global, optical transport networks. They include such heavyweights as Qwest, Level 3, Global Crossing, 360Networks, and Tyco. Their intention has always been to make bandwidth as inexpensive as possible in order to become the *carrier's carrier*, selling bandwidth to everyone in huge, inexpensive quantities. Most of them started as commodity providers, but soon moved up the optical food chain as they added new services. Consider, for example, Qwest CyberSolutions, the joint venture company between Qwest and KPMG that provides global ASP services. Qwest's network architecture speaks to their commitment to provision massive amounts of bandwidth: two diversely routed conduits house two cables, each with 96 optical strands per cable. Each strand is further subdivided through DWDM into 40 individual wavelengths, each operating at OC-192; the result is a network that offers an almost unimaginable 154 Tbps of bandwidth, many times the capacity of AT&T's network. Again, the key to success in this market is to provide more than simply bandwidth.

The bandwidth barons stand to benefit from the realization of the optical network edge for many of the same reasons already stated: it enables them to reduce the complexity of their networks, thus influencing the speed and quality with which they can deploy service.

The Internet Service Providers (ISPs)

The ISP, particularly the facilities-based ISP, typically builds a network by leasing high-bandwidth optical facilities for the interconnection of their routers. To them, the network edge lies at the metropolitan carrier point-of-presence where the ISP's edge routers reside. Their biggest concern today is the incredibly competitive nature of their business and ongoing winnowing of the ISP herd.

ISPs benefit from the creation of the optical edge because it enables them to expand the variety of their IP-based services, thus providing much-needed differentiation.

Connecting to the Last Mile

Technology has one universal truth: the closer one gets to the end user, the more diverse and complex the network becomes. In the world of optical networking, this statement is equally true. At the network's edge, the local loop must support multiple access technologies and multiple standards, including SONET, SDH, frame relay, ATM, and IP. In response, companies have emerged that focus on the creation of wide spectrum network management systems that not only handle multiple protocols and services, but are flexible, scalable, single-seat systems that reside at the periphery of the network, close to the customer, rather than in the shadowy recesses of the central office. These systems will support the simultaneous provisioning of broadband voice and data applications, and will help to realize the true promise of convergence.

So will fiber ever reach the home or small office? Absolutely. However, certain caveats must first be satisfied, such as the cost of optoelectronics, the dearth of optical interfaces on consumer equipment, network limitations, and regulatory issues. In the meantime, alternative solutions have emerged with varying degrees of success, such as *Hybrid Fiber/Coax* (HFC) architectures that take advantage of in-place, fully functional wiring schemes. HFC is considered to be a relatively low-cost solution for extending the reach of fiber. In the last few years, however, a number of fiber-to-the-home projects have been initiated worldwide, providing optical local loop connectivity to more than 300 million potential lines.

Metropolitan access is characterized by the deployment of ring architectures, used for the aggregation and transport of low-speed traffic. For example, a carrier might deploy a 10-Gbps SONET or SDH metropolitan ring throughout a large city, which would then interconnect to lower speed, 2.5-Gbps access facilities—either point-to-point circuits or rings. SONET/SDH, as well as DWDM, are key technologies in this environment; ILECs/Incumbent PTTs and CLECs/City Carriers are involved in this segment of the marketplace, where they wish to serve as peering points in the network, providing high-speed interfaces to multiple protocols, technologies, and companies.

The ILECs/Incumbent PTTs, on the other hand, face unfettered competition from all sides, but they remain in an enviable position: they control access to the bulk of the customers through their control of the access lines. Their primary goal is to add high-speed access as quickly as they possibly can while continuing to ensure that they can deliver on promises of QoS.

Some service providers have reinvented themselves as broadband access carriers, offering a wide array of high-bandwidth access options, including DSL, cable modems, and wireless solutions. These broadband access providers face a different set of issues than traditional carriers. First, they do not have a great deal of experience managing high-bandwidth access services, and have never seen the tremendous growth that currently characterizes the market. Second, they tend to be quicker and more nimble than their traditional counterparts, making technology decisions that are in the best interest of their customers based on economies of scale and the potential to generate added revenue in innovative ways. These companies must deploy the most current technologies and must ensure scalability if they are to meet the growing demands of their customer base. Many believe that these companies may eventually "own" the customer as broadband access catches on. They are deploying high-speed architectures designed to support the requirements of telecommuters, remote office installations, wireless business access, Gigabit Ethernet interoffice communications, and regional ISPs. For example, a carrier might deploy an OC-48/STM-16 metro ring that provides transport for traffic that originates on DSL-equipped local loops, interconnecting remote workers with a corporate network elsewhere in the metro area. Another carrier might deploy an all-optical metro infrastructure to support the huge traffic increases between base stations that result from the deployment of third-generation wireless services. As data access over cable becomes more common, a carrier might build a high-speed access network designed to aggregate and transport traffic between cable customers and an ATM backbone.

As wireless local loop technologies such as the *Local Multipoint Distribution Service* (LMDS) find their way into the business access domain, broadband metro carriers will roll out high-speed rings to satisfy the demands of these and other similar services. Similarly, a regional ISP, with a need to connect to multiple *Network Access Points* (NAPs) and database locations, may be served by a broadband metro carrier's 10-Gbps metropolitan ring, which provides transport for traffic that originates on low-speed services, such as OC-48/STM-16, OC-3/STM-1, and traditional TDM services. Finally, a metropolitan ring can be used to interconnect corporate LANs; several vendors have deployed multiprotocol multiplexers that enable Gigabit Ethernet to be transported across a 10-Gbps ring—clearly an application with promise given the widespread deployment of Ethernet technology.

Metropolitan transport is precisely what the name implies: the segment of the transport market that delivers the high-speed rings used to aggregate and move low-speed traffic between locations or onward to a wide area transport environment. Finally, metropolitan enterprise is the realm of innovative access techniques designed to provide high-bandwidth solutions for businesses.

Reinventing the Network

Contrary to the belief that many would like to have, SONET and SDH are far from dead. They are among the most widely deployed transmission systems, and are deeply embedded in long distance, enterprise access, and even metro transport and access networks. The current direction seems to indicate that these technologies will enjoy a long, slow decline to retirement as more capable optical technologies replace them. However, this will take the form of a slow evolution, not a slash-cut revolution. As the demand grows for more flexible bandwidth allocation, the technology will see to it that the available bandwidth exceeds the demand. The network will grow and evolve to accommodate the changes in the user and application profiles, creating enormous opportunities for emerging optical transmission technologies. The downside, of course, is that this evolution brings with it a rise in complexity as additional network elements are added, placing greater demands on network management systems to ensure that they are up to the task of meeting the unwavering demands of an increasingly exacting and technically competent customer.

Toward a New Network Paradigm

The answer to this conundrum is to reduce the overall complexity of the network by lowering the number of network elements, replacing copper networks with optical, and improving the reliability of the network as a whole. These can be done, but in some cases, they run contrary to the direction that the network has taken. For example, until recently, optical components were corralled in the physical layer as SONET or SDH devices, providing physical transport capability for ATM and frame relay networks, which in turn provided switching fabric for Layer 3 protocols, such as IP. With the arrival of *wavelength division multiplexing* (WDM) in the late 1990s, an additional optical sublayer was created below SONET/SDH, and at the end of the 1990s, we saw the arrival of optical switching, which burrowed into the space between the physical layer (SONET/SDH) and the switching layer (ATM/frame relay). This resulted in a growth in complexity as the three-layer protocol stack became a five-layer protocol.

Today, a move is afoot to collapse the stack again by evolving to an all-optical network. Current product offerings make it possible to deploy optical networks not only in the long haul, but in the metro and access regions as well.

Why the Evolution?

One of the principal driving forces behind the evolution to all-IP networks is *convergence*. Traditionally defined as a technology phenomenon, convergence has more recently been recognized as a troika comprising a technology component, a company component, and a services component. *Technology convergence* defines the vortex that is inexorably pulling the industry toward packet-based IP networks because it makes sense to do so. As *quality of service* (QoS) protocols and techniques continue to evolve, and as network devices evolve to be able to respond to QoS demands, services will move to the IP network because it provides a quality-centric, universal protocol substrate that will support all service types across a single network infrastructure. From the point of view of a service provider, this is an ideal construct because it means that they can eliminate the multiple networks, support services, and billing infrastructures that they currently operate in favor of a single converged transport environment. Today, the typical large service provider operates the *public-switched telephone network* (PSTN), the frame relay network, the ATM network, the IP network, the ISDN and DSL overlay networks, and in some cases, wireless and X.25 networks. For each of these, they must have maintenance, installation, operations, accounting, configuration, and provisioning systems, and personnel and network management centers capable of monitoring them. The fact that they communicate with each other and occasionally share resources and customers makes their management that much more complex.

Now, consider the advantage of the IP migration that is inexorably underway. If convergence is real and it becomes possible to migrate all of these disparate services to a single network infrastructure, the service provider will realize significant benefits as the business of providing network services undergoes massive simplification. Already, ATM providers offer a service called *frame relay bearer service* (FRBS) from their ATM networks; customers requesting frame relay service are handed a frame relay interface, and their frames are transported into the network. There, they are converted to cells for wide area transport across the ATM backbone, and converted once again to frames upon delivery to the receiving interface. Does the customer know that his or her data was actually transported as ATM traffic? No. Does it matter? Absolutely not—provided the agreed-upon quality of service stipulated in the frame relay service level agreement is met. In this example, the service provider has the ability to essentially halve their network complexity by delivering two services from a single network.

If we take this to its ultimate conclusion, we find a network that is capable of delivering all services with equally appropriate QoS levels. The service provider truly has become a *service provider*, now capable of focusing on the demands of the customer rather than on the underlying technology.

However, the convergence phenomenon involves more than just IP implementation and the other attendant technologies that make possible the migration to a packet-based infrastructure. A second aspect of convergence is *company convergence,* defined by the merger and acquisition feeding frenzy that is currently underway in the telecommunications industry. Most companies have realized that although they are technologically very good at what they do, two other critical truths define them. First, more and more, customers don't care, and don't *want* to care, about the underlying technologies that make their services work. This is illustrated in Figure 6-19. They are content to simply accept the services that the technologies make possible and happy to know that the service provider will worry about the network, leaving them to worry about their own products and services. Second, as customer demands become more stringent, players in the telecommunications game have come to realize that they need more capability than they currently have. Given the pace of the industry they are in, they also know that they do not have the time to develop that capability in-house, so instead, they go out and buy it in the form of an acquisition, a merger, or a strategic alliance. Their overall set of capabilities, then, grows with minimal pain and at a pace that is acceptable to the customer.

The third component of the convergence troika is *services convergence*. Tied in tightly to the other two parts, service convergence speaks to the fact that the average customer is looking today for a single source for their telecommunications networking services. The fact that those services may derive from the combined capabilities of multiple smaller companies is immaterial; what matters is that the customer gets the products and services that he or she requires to do business.

How does this tie together, then? Rather elegantly, actually. If a product or service provider properly assesses the needs of their customers, they can position themselves to be much more than simply a box provider.

Consider the following example. Acme Router Corporation sells high-quality hardware that supports the growing Internet industry. When its

Figure 6-19
The customer's (ideal)
view of the network.

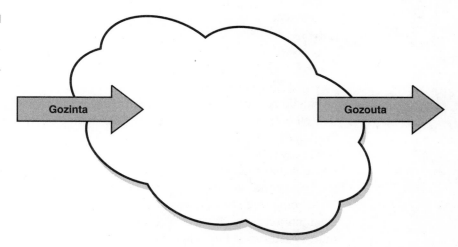

Gozinta

Gozouta

customers call to place orders, Acme responds quickly, shipping precisely what the customer asks for, always within the time frame requested. Acme doesn't really know why they are called, but they are happy to be the vendor of choice. In this case, Acme is nothing more than a commodity provider. Chances are good that they are being chosen because they are the lowest-cost vendor, and because commodities are defined as products that are differentiable on price alone; this is probably not the best business model for Acme to engage in.

Now, change the model slightly. Acme, now a proactive provider of router-based services, expends considerable effort to understand not only the business drivers of their own customers, but also the forces that affect their *customers'* customers. By understanding what it is that drives the so-called third tier, Acme can anticipate the requests that its direct customers will have, and provide a solution when they call—or better yet, be able to call *them* with a proposal based on Acme's unique understanding of the customer's marketplace. If Acme understands what services the third tier will be looking for, then they understand what services they should be positioned to provide. Furthermore, if they know the services they should have in their pantheon of capabilities, then they know what technologies they must add to their collection to meet those requirements, and therefore know which technology companies they must ally with in order to round out their complete service offering. This is convergence at its best: a combination of technological capability, marketplace understanding, and an unrelenting focus on service with an understanding that it is service, not technology, which drives the telecommunications economy.

The Evolving IP Model

As the traditional voice-centric central office bends to the pressure of diverse quality of service requirements from heterogeneous traffic types, a new network architecture is evolving that satisfies these demands. The architecture, shown in Figure 6-20, comprises a four-layer protocol stack. The *Internet Protocol* (IP), shown at the top, lies at the center of this great evolution. With the possible exception of the SS7 signaling protocols, IP is the world's most widely deployed protocol. It also provides us with the only truly universal addressing scheme in existence, and is embedded deeply in every network operating system deployed today that is of any consequence. It has become the focal point for such environments as call centers, where universal routing of multiple traffic types to a single operator is highly

Figure 6-20
The new network model.

desirable. Prior to the introduction of IP and unified messaging, the ability to do that was complex, costly, and people intensive. Each operator required multiple phone lines, and the call center required call routing software that was complex and costly in its own right. IP, in concert with other Internet-derived protocols such as HTTP, allows for tremendous simplification of the call routing algorithm. Consider the following example. To contact the author of this book using every possible business contact technique, you would need an office telephone number, a fax number, a home telephone number, an e-mail address, a cell phone number, a pager number, and so on. Chances are very good that the numbers would not be sequential and would therefore be difficult to remember.

IP's Promise

Now, consider the promise of IP's unified messaging concept. Instead of multiple unrelated numbers, the author could be contacted over an IP network in every possible way by typing

- **Steve@office.ShepardComm.com**
- **Steve@fax.ShepardComm.com**
- **Steve@home.ShepardComm.com**

- Steve@mail.ShepardComm.com
- Steve@cell.ShepardComm.com
- Steve@pager.ShepardComm.com

Clearly, this is a far simpler way to operate a business or call center, and dramatically simplifies the contact process for the customer. IP, then, provides the global addressing and universality required to make this possible.

Today, IP has a single drawback that is something of a showstopper. Because it was originally designed for the routing of connectionless, delay-insensitive data traffic across a packet network, it does not provide adequate QoS granularity to the broad range of services that it is now being asked to deliver. IP is something of a proletarian protocol in that it treats all traffic equally. By and large, in its *native mode*, it is incapable of discriminating between high- and low-priority packets. This, of course, is a problem because the diverse nature of traffic today requires a variety of QoS levels if the service provided by this single network fabric is to sell. Several options are either available or under development to accomplish this.

IP Version 6 (Ipv6)

The first of these is the next generation of IP, known as *IP Version 6* (Ipv6). In Ipv6, the protocol header has been redesigned to provide space for specific bytes that can be used to indicate the QoS parameters required for each packet so that network routers can handle them accordingly. However, IPv6 is far from ready to be commercially deployed, and although it has been tested and is being trailed today, its widespread deployment is still a bit over the horizon.

Tag Switching

A second method is to use a technique called *tag switching*. Originally developed by Cisco for quality control in large router networks, tag switching precedes each packet with an additional field, called a *tag*, which contains QoS requirements that network routers can take into account as they make routing decisions. Tag switching is a very capable technique, but has the drawback of being proprietary—it only works on Cisco routers. In response, an open, vendor-independent form of tag switching was developed called *Multiprotocol Label Switching* (MPLS).

Multiprotocol Label Switching (MPLS)   When establishing connections over an IP network, it is critical to manage traffic queues to ensure the proper treatment of packets that come from delay-sensitive services such as voice and video. In order to do this, packets must be differentiable, that is, identifiable so that they can be classified properly. Routers, in turn, must be able to respond properly to delay-sensitive traffic by implementing queue management processes. This requires that routers establish both normal and high-priority queues, and handle the traffic found in high-priority routing queues faster than the arrival rate of the traffic.

MPLS delivers QoS by establishing virtual circuits known as *Label Switched Paths* (LSPs), which are built around traffic-specific QoS requirements. Thus, a router can establish LSPs with explicit QoS capabilities and route packets to those LSPs as required, guaranteeing the delay that a particular flow encounters on an end-to-end basis. It's interesting to note that some industry analysts have compared MPLS LSPs to the trunks established in the voice environment.

MPLS uses a two-part process for traffic differentiation and routing. First, it divides the packets into *Forwarding Equivalence Classes* (FECs) based on their QoS requirements, and then maps the FECs to their next hop point. This process is performed at the point of ingress at the edge of the network. Each FEC is given a fixed-length *label* that accompanies each packet from hop to hop; at each router, the FEC label is examined and used to route the packet to the next hop point, where it is assigned a new label.

MPLS is a *shim* protocol that works closely with IP to help it deliver on QoS guarantees. Its implementation will enable the eventual dismissal of ATM as a required layer in the multimedia network protocol stack. Although it offers a promising solution, its widespread deployment is still a ways in the future because of the logistics of deployment.

Multiprotocol Lambda Switching (MPλS)   *Multiprotocol Lambda Switching* (MPλS) is the latest innovation to come along in some time. Lambda switching (sometimes called photonic or wavelength switching) is used in optical networking to switch individual wavelengths onto separate paths for specific routing. In conjunction with technologies such as DWDM, which enables 80 or more separate wavelengths to be transmitted on a single optical fiber, lambda switching enables a light path to behave like a traditional virtual circuit.

Lambda switching works in much the same way as traditional routing and switching. Lambda routers, which are also called wavelength routers or optical cross-connects, are positioned at network junction points. The lambda router takes in a single wavelength of light from a fiber and recombines it into another fiber. Lambda routers are being manufactured by a number of companies, including Ciena, Lucent, and Nortel.

Multiprotocol Lambda Switching is a variation on the MPLS theme, where specific wavelengths are used instead of labels as circuit identifiers. The specified wavelengths, like the labels, make it possible for routers and switches to perform routing functions without having to open the packet for addressing information.

Asynchronous Transfer Mode (ATM)

The third technique, and the one that holds the greatest promise today, is found one layer down the protocol stack. ATM provides granular QoS control through the capabilities of its *Adaptation Layer* (AAL).

The ATM Adaptation Layer at the ingress switch examines customer traffic as it arrives at the switch and then, based on the nature of the traffic, classifies it according to its QoS requirements based on three parameters: whether the traffic is connectionless or connection-oriented, whether it requires a fixed or variable bit rate, and whether or not an explicit timing relationship exists between the sending and receiving devices. Once these have been determined, the ATM ingress switch assigns a service class to the cells that make up the traffic stream and transmits them into the network, knowing that ATM's highly-reliable connection-oriented transport architecture and each switch's capability to interpret and respond to the assigned service class will ensure that the QoS mandate of the sending device will be accommodated on a network-wide basis. Thus, ATM provides QoS today.

Dense Wavelength Division Mulitplexing (DWDM)

Finally, we arrive at the lowest level of the protocol stack where we encounter DWDM. DWDM offers massive bandwidth multiplication capability and is in widespread use today. DWDM is discussed in more detail elsewhere in the book, but suffice it to say that it is a form of frequency-division multiplexing, operating in the infrared domain, that enables multiple wavelengths of light to be simultaneously transmitted down the same fiber, significantly increasing the available bandwidth of the fiber and providing a cost-effective bandwidth multiplication solution to the provider. In fact, industry estimates show that the per-mile cost to trench in new fiber as a bandwidth relief effort is approximately $70K per mile; adding the same bandwidth by changing the end points to include DWDM costs a much-reduced $12–20K per mile.

There is nothing new about DWDM. In fact, it relies on technology that has been in widespread use throughout the network since the 1960s in the form of frequency-division multiplexing, the technique of dividing a broad swath of spectrum into chunks and assigning each chunk to a different customer. As a technique for the facile multiplication of bandwidth, DWDM is a technological hero.

Of course, the success of WDM is more involved than simple multiplexing. Fundamental to its success was the capability to eliminate the *optical-to-electrical-to-optical* (O-E-O) conversion that was necessary for switching, multiplexing, and signal regeneration—a process analogous to amplification in the analog transmission world. The first important accomplishment that led to this simplification was the development and widespread deployment of the all-optical amplifier.

Optical amplification, explained earlier, is the direct result of a sublime understanding of quantum physics. *Erbium-Doped Fiber Amplifiers* (EDFA) amplify signals in the optical domain, completely eliminating the O-E-O conversion that must normally take place.

EDFA is the technology that makes Wavelength Division Multiplexing commercially possible. By eliminating the need for electrical to optical conversion, the promise of the all-optical network can begin to be realized.

Before WDM became commercially available, optical transmission systems were for the most part limited to the transmission of a single wavelength per fiber, thus limiting the bandwidth of that fiber rather substantially by today's measure. As optical networking techniques continued to advance, however, this limitation became a non-issue. The original WDM systems developed by Lucent Technologies' Bell Laboratories had the capability to transmit as many as four wavelengths of light down a single fiber. Today, DWDM systems routinely carry as many as 160 different wavelengths per fiber by assigning a different frequency, or color, to each stream of information. Individual lasers operating at different wavelengths, or single lasers operating at multiple wavelengths, transmit the information into the fiber, thus enabling enormous bandwidth to be offered from a single fiber.

At the time of transmission, the optical signal is amplified at the ingress point, after which it enters the fiber. Depending upon the nature of the fiber itself, the signal is then amplified every 40 to 60 miles to overcome the inevitable weakening of the signal that occurs over distance.

Using DWDM, service providers have the ability to add bandwidth to an existing fiber-optic network without having to engage the services of a backhoe. This becomes important when the costs involved are examined. As mentioned earlier, when adding fiber to an existing network as a way to increase available bandwidth, the cost can be as much as $70K per mile. On the other hand, the addition of DWDM electronics at the end points to accomplish the same thing can cost as little as $12–20K—a significant difference. Much of the cost, of course, is labor, the requirement for which is dramatically reduced when the need for outside plant work is eliminated. This technique is often referred to in the industry as the deployment of *virtual bandwidth* because physical resources have not been added to bring about the improvement in the network that has taken place. In the same way that EDFA provided low-cost and highly effective amplification to optical spans, other innovations have helped to reduce the complexity and expense of the migration from electrical to hybrid to all-optical networks. For example, optical switches, which use arrays of micro-mirrors, refractive bubbles in fluid-filled chambers, and the natural resonant frequency of certain types of crystals, make it possible to eliminate electrical switching elements. Tunable lasers eliminate the need for multiple lasers operating at specific frequencies and reduce sparing requirements for service providers. Advanced and highly accurate optical filters provide channel separation in densely packed wavelength division systems, making it possible to dramatically expand the total bandwidth of an optical span. Management and monitoring systems, specifically designed for optical networks, make it possible to discretely manage these networks at highly granular levels, thus ensuring the ability to meet the requirements of customer service level agreements.

Protocol Assemblies: Putting it Together

IP, ATM, SONET/SDH, and DWDM represent a powerful and robust protocol stack, but in the minds of many industry pundits, they are far from being the ultimate network design for the full-service transport fabric. One significant complaint that is often voiced about this four-tier stack is that it is highly overhead intensive. True enough: IP, ATM, and SONET/SDH all add considerable overhead in the process of doing what they do. However, remember the adage: "if it ain't broke, don't fix it." Although this phrase has some truth, many argue that the model *is* broken and can be significantly improved.

Consider what happens to customer traffic as it enters the following IP-based network. The stream of data enters the ingress router where it is chopped into pieces. A header is attached to each piece, which contains information used to route the packets from the source to the destination. This header, although substantial, is usually inconsequential because IP packets tend to contain thousands of bytes of payload (user data).

The IP packets are then handed down to the ATM layer, where they are further segmented into 48-octet pieces. Each is given a five-byte header to form ATM cells. Now, the overhead in the header becomes significant: approximately 10 percent of the cell is overhead.

The cells are then handed down to the SONET/SDH layer, where they are packaged in frames for transport across the optical network. Each frame has embedded within it additional overhead, to the tune of about 5 percent of the frame.

It should be clear to the reader that this four-layer stack has some rather serious downsides. First of all, IP, as it exists today, although a good protocol for universal networking, does precious little to guarantee the integrity of the user's data. ATM, for all its capabilities, is not really ideal for anything. It is not the best scheme for the transport of voice; the PSTN has it beat hands-down. It's also not the best for video; a dedicated high-speed circuit is far better. It certainly isn't the best solution for data transport; it's far too expensive, and other solutions are equally capable.

Furthermore, the overhead tax that IP, ATM, and SONET/SDH exact is significant. Many argue that it doesn't matter because of the belief that we are entering a time when bandwidth will be so abundant that we can afford to waste it. However, building networks based on that belief is irresponsible and dangerous. The communications corollary to Parkinson's Law promises that we will find a reason to need that bandwidth, so exercising caution to be efficient is advice worth listening to. Service providers are already burdened with the legacy of SONET and SDH, which were designed in a time when those deploying them were monopolies and not terribly concerned with protocol efficiencies. Many find the two to be monolithic, overhead-intensive, and inefficient—a stand that is hard to argue with.

One effort that is afoot (and that will undoubtedly be successful) intends to collapse the four-layer stack to two, shown in Figure 6-21, eliminating the ATM and SONET/SDH layers entirely by moving their responsibilities into the IP and DWDM layers, respectively. In other words, the responsibility for QoS control would be moved upward to IP, whereas survivability and robustness would become the responsibility of DWDM. To accomplish this,

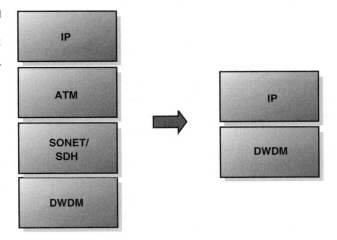

Figure 6-21
The evolving network model.

several things must happen. First of all, IP must become capable of managing and carrying out the guarantee of QoS, without the benefit of ATM's adaptation layer. This could be done in several ways. First of all, the *Type of Service* (TOS) bits in the IP header could be used in concert with the *Differentiated Services* (DiffServ) protocol to create and respond to multiple QoS levels. This would require universal implementation of DiffServ/TOS throughout the greater network.

Second, MPLS or MPλS could be deployed. This is the current favorite solution.

Finally, some network stalwarts believe that IPv6 will arrive and be widely implemented, making possible the very granular QoS management that its overhead makes possible. One way or another, the function will be migrated upward.

Equally important is the migration of SONET/SDH's responsibilities to guarantee survivability downward into the DWDM sublayer. Today, DWDM provides massive amounts of bandwidth through the expediency of frequency-division multiplexing. By dividing the optical bandwidth into channels, total throughput can be multiplied many times over. However, native-mode DWDM does nothing to guarantee the integrity of the information that it transports; it serves as nothing more than a multistream fire hose. Consequently, SONET/SDH's automatic protection switching, self-healing ring support, and embedded network management protocols are necessary if the integrity of the network is to be guaranteed.

A new technology will soon change that requirement. Known generically as Digital Wrapper, it encloses each wavelength's traffic in a low-overhead frame of additional data that enables DWDM to detect and correct errors using sophisticated forward error correction techniques, perform optical layer performance monitoring, and provide ring protection on a wavelength-by-wavelength basis. Thus, many of the functions traditionally provided by SONET/SDH will be assumed by DWDM, eliminating the requirement for yet another protocol layer. Thus, the collapse of the four-layer protocol model into a two-layer construct is possible and highly likely. Already, carriers like Yipes! and Telseon are providing Ethernet transport across optical networks with significant success, particularly in the metro area. Others will certainly follow.

The result of this protocol evolution is that functions traditionally performed within the core of the network—aggregation, prioritization, policy enforcement, QoS, and concentration of traffic—can now be performed at the edge of the network in the customer provided equipment.

A final responsibility that is migrating from the core to the edge is switching (and routing), as shown in Figure 6-22. Historically, this process has been centralized because the devices required to do it were large and inordinately costly. Today, thanks largely to advances in microelectronics, that process, as well as the signaling responsibilities that govern it, is being substantially moved into high-speed edge routers.

So what's left in the core? In reality, the only thing left is very high-speed transport; all setup processing, QoS deliberation, traffic discrimination, and concentration are now performed by intelligent edge devices, whereas the core is left to provide extremely high-bandwidth transport, as shown in the illustration. This is the domain of optical networking. This division of

Figure 6-22
From core to edge.

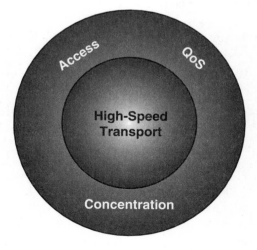

labor makes a great deal of sense and is guiding the design of modern high-speed networks as they struggle to meet the growing demands of bandwidth-hungry applications.

As the focus migrates from the core to the edge, one truth emerges: there is enormous interest in the access region, and metropolitan transport is the hottest commodity available. Referring back to the network skin effect described earlier, a lot of surface area is available to touch at the edge, and companies are rising up to do it. Optical networking is moving into all three regions of the metro space in a big way: metro core, metro access, and enterprise.

Next-Generation SONET and SDH: What's Next?

The fastest growing region of the optical networking world is the network edge, particularly the optical edge. As SONET and SDH evolve to meet the burgeoning demands of an ever-evolving marketplace, certain requirements become clear and rather obvious. Let's start with the most rudimentary of them. The vast majority of SONET and SDH add-drop multiplexers do not have data interfaces. This means that data traffic, which tends to be relatively efficient, must first be packaged into ATM cells or frame relay frames, or transported across a legacy T1 or E1 facility with all its inherent inefficiencies. The result of this, in addition to cost, is the need for added managerial capability and increased infrastructure complexity.

If we consider the characteristics of SONET and SDH that the technologies are known for and attempt to prognosticate about their future, we see some potentially interesting trends developing that will undoubtedly affect (albeit positively) the long-term potential of the two technologies.

The first of these is management, often called OAM&P or OAM. Today, the SONET and SDH's inherent capability to provide network management capability, and to therefore guarantee to the degree possible the survivability of the network, is without question their greatest and most strategically important selling point. Over time, however, as demand shifts and the desire to reduce protocol overhead achieves the frenzy stage, the SONET and SDH overhead may well be traded to one degree or another in favor of less overhead-intensive options, such as Lucent Technologies' WaveWrapper. WaveWrapper is a digital wrapper solution that *wraps* optical channels (see DWDM) in an overhead wrapper that permits individual optical channels, or lambdas, to be monitored for transmission quality and error control. Similarly, network providers are flirting with a collection of new protocols that provide quality of service control across the wide area, including multiprotocol label switching (MPLS) and multiprotocol lambda switching (MPλS).

The next item that appears on our spontaneous list is the ring architecture itself that both SONET and SDH are known for. Today, rings provide the critical capability for mission-critical data networks to survive node and fiber failures. They are therefore used universally. In the future, however, rings will undoubtedly disappear from the long-haul environment in favor of mesh topologies; they will, however, remain as viable options in the metro domain primarily for aggregation purposes.

The functional interplay that exists between the electrical and optical worlds, long a part of the SONET/SDH game, will slowly diminish as the all-optical network begins to grow. Already we have seen a slowing in the sale of add-drop multiplexers as all-optical switching has emerged on the scene, although this has not hurt manufacturers as the average price of these devices has climbed. Similarly, the O-E-O conversions that take place at the electrical ports on ADMs and *Terminal Multiplexers* (TMs) will slowly be phased out as manufacturers warm to the idea of the all-optical switch and begin to design and sell them.

Finally, the SONET and SDH overhead, particularly the framing bytes, will continue to be important. As alternative Layer 2 protocols such as Ethernet grow in popularity for use in the metro marketplace, however, we could see a gradual diminishment in use of SONET and SDH.

Next on the list is efficiency. Because SONET and SDH rings typically reserve half their bandwidth for protection purposes, many service providers are interested in upgrading network infrastructures to overcome this perceived inefficiency. This reservation technique is fine for the legacy requirements of circuit-switched voice, but because data networks typically have their own methods for dealing with network failures or severe delay, traditional SONET or SDH protection is often perceived as overkill.

Scalability appears next. SONET and SDH tend to be "all or nothing" systems in which the need to add capacity at a single node on a ring requires an upgrade to the entire ring, regardless of the growth requirements at the other nodes. In the wide area, this is less of a problem because traffic requirements tend to be comparatively homogeneous. In the metro area, however, where node-by-node demand tends to be highly variable and unpredictable, the impact is greater because of the inherently more complex provisioning and management concerns that crop up.

Equally problematic is the inherently data-hostile nature of the legacy transport schemes around which SONET and SDH are built. Because the framing structures are highly regimented and inflexible, bandwidth upgrades tend to occur in large, inefficient blocks. This leads to cost concerns and wasted bandwidth.

Next-Generation SONET and SDH

These issues, all of which are valid, have led to talk about the so-called next-generation SONET or SDH network. This model, based on work performed by a variety of vendors, including Astral Point, Mayan, Cisco, Lucent, Nortel, Redback, Sycamore, Fujitsu, Corvis, Ciena, and a host of others, is based on a structure in which data is mapped efficiently into SONET or SDH transmission streams such as virtual tributaries or channelized and unchannelized optical streams based on either existing standards or proprietary, but widely accepted mapping schemes that guarantee interoperability. Cisco has been particularly active in this area and others. Their 15454 system, for example, delivers wideband packet services over SONET, using multiple bonded virtual tributaries (VT1.5s, specifically) to create point-to-point circuits that satisfy the specific bandwidth demands of various data types. For the bitheads in the audience, their work is based on two principal standards: RFC 1619, which addresses the use of *point-to-point protocol* (PPP), and RFC 1990, which concerns itself with *multilink PPP* (MLPPP) for encapsulation of data traffic into DS-1 payloads.

Given the direction that the overall industry (including user communities) is going, the following model seems likely for the overall design of next-generation SONET and SDH networks. Data, probably IP that originates in a corporate environment and is most likely framed as some form of Ethernet, passes from an Ethernet port across one of the previously mentioned PPP interfaces into the SONET or SDH network. There, the data is mapped into framed DS-1 or E-1 channels, which in turn are prepared for transmission over a ring.

As was previously mentioned, a number of vendors have devoted considerable resources to the development of next-generation SONET and SDH products. They will be described in detail later in this section.

So what is the addressable market for the next generation of these technologies? They are largely the segments that we described earlier. These include the central offices owned and operated by both incumbent and competitive carriers in metropolitan areas, particularly those that serve customer bases that will use high-bandwidth access technologies such as DSL,

IP, ATM, broadband wireless, and a variety of others; the ISP points-of-presence where traffic aggregation is the rule and multiple network operators congregate for access; interexchange carrier offices, where the variety of service types are best served at the network edge; and Internet data centers, where the rule of the road is massive, highly flexible bandwidth.

All of these are factors that strongly affect the viability of SONET and SDH in the general marketplace, particularly at the network edge for data applications.

So, where are SONET and SDH going? The correct answer is nowhere fast—and that's a good thing. A variety of factors ensure their survival for the foreseeable future. These include the tangible growth of next-generation SONET and SDH hardware systems from capable, devoted-to-the-market vendors, an enormous installed base of legacy networks, and a significant level of comfort on the part of the carriers with traditional SONET and SDH technologies. Add to these the growing demand for high-speed interfaces, the seemingly unstoppable demand for bandwidth, and the almost hysterical frenzy that characterizes the carriers' attempts to expand their service offerings as they battle for market position. In North America alone, the growth of new entrants (CLECs, bandwidth barons, ISPs, and others) threatens to push the SONET market from a respectable $930 million in 2000 to an almost unbelievable $6.3 billion by 2004. This does not sound like the behavior of a dying legacy technology, does it?

The Equipment Market

The jockeying for position that continues in the SONET and SDH markets is a constantly evolving process. In a recently published report, consultancy RHK awarded the number one marketplace position in the lucrative OC-48 long-haul segment to Cisco, with marketshare rocketing 500 percent from 2.3 percent of the market in 1999 to a whopping 11.3 percent in 2000. Other vendors enjoy equally impressive positions: of the $1.9 billion spent on long-haul OC-48 SONET equipment in 2000, Cisco enjoyed 30 percent of it, Nortel garnered 27 percent, NEC 21 percent, Lucent Technologies 16 percent, Alcatel 3 percent, Hitachi 1 percent, and the remaining 2 percent was shared among a variety of others. Those numbers change dramatically, however, when we focus instead on the fastest growing segment: the 10-Gbps market. There, Nortel is the one to beat with 95 percent, whereas Alcatel and Fujitsu provide a distant second at 2 percent each.

The North American metro market is equally dynamic. According to RHK, Fujitsu is the leader with 43 percent of the pie, with Lucent as the clear second place player at 23 percent. Nortel has 22 percent, Cisco 8 percent, Alcatel 2 percent, and NEC 1 percent.

Globally, this technology continues to run rampant. According to Dell'Oro Group, global SONET sales exceeded $17 billion in 2000, and are expected to continue their run-up throughout 2001. Let's now turn our attention to these companies.

ADC

It is commonly known that traditional TDM-based SONET and SDH systems were originally designed for voice. Today's Internet-driven traffic growth, however, consists primarily of data. The challenge that carriers face is to provide voice services while continuing to manage exponential growth of data traffic. ADC's solution to this challenge is the Cellworx™ *Service Transport Node* (STN).

The Cellworx STN supports voice services, video, data, and Internet traffic, all in the same system. The architecture is scalable and supports rates from T1/E1 to OC-48c/STM-16c. The Cellworx STN is unique in that it enables carriers to provision exactly the amount of bandwidth the customer requires and transport only the bandwidth used.

Appian

Appian has an interesting approach: they combine packet switching and TDM transport to forge a system that offers services somewhere between those of the metro environment and the next-generation systems approach. Appian's strategy is to add the flexibility of a packet network to the tremendous capabilities of a SONET/SDH network through the deployment of a fully SONET and SDH-interoperable architecture. Known as Optical Services Activation Platform™, Appian relies on four key capabilities to achieve this goal: shared payloads and paths, packet QoS, Optical Data Protection™, and an Ethernet service interface. Shared payloads and paths is a distributed packet-switched platform that enables traffic from a variety of sources to share a SONET or SDH payload to a common network destination, thus offering significantly better bandwidth utilization. Packet QoS ensures the ability to offer a wide variety of QoS levels by provisioning separate queues for each of four priority levels for each service, adjusting queue depths based on varying priority and bandwidth requirements, dynamically sharing buffers, utilizing multiple disparate congestion control methodologies, and deploying multiple classes of service. Optical Data Protection provides 50 ms recovery using a technique that is similar to SONET line switching. Finally, the Ethernet service interface eliminates the cost and complexity of added routing functions in corporate networks and provides three services: Ethernet-to-Ethernet Access, Ethernet Private Line, and Ethernet Virtual Network Services.

Astral Point

Astral Point's Optical Node 5000 is designed to support both ring and mesh architectures and is focused primarily on the metro market where hybrid installations shine. Astral Point enjoys success with three major customers: Lighthouse Communications, Time Warner Telecom, and Advanced Telcom Group; all three of which have installed Astral Point's ON 5000 solution for integrated connectivity.

Atmosphere Networks

Atmosphere Network's NTU 300 is a customer-based device that aggregates T1, Ethernet, and ATM traffic. These combined flows are then piped to Atmosphere's Full Service Node, the FSN 1200, which is installed at the service provider's central office.

Ciena

Ciena, with its recent contract announcement to deliver a wide area DWDM-equipped mesh network to CLEC McLeod USA, is clearly targeting that segment of the market. Ciena recently acquired Omnia, which nicely rounds out their capability suite.

The company's MultiWave CoreDirector™ is an intelligent optical networking core switch that is designed to deliver differentiable end-to-end optical capacity across the network, with flexible protection options. CoreDirector features the networking intelligence of Ciena's LightWorks

Operating System, which permits end-to-end optical provisioning and management in addition to highly capable protection. With its scalability, flexibility, and advanced networking options, CoreDirector reduces the cost of deployment, operations, and scalability.

CoreDirector is capable of delivering 640 Gbps of non-blocking, bidirectional switching, with the capability to upgrade to 38 Tbps. It supports 256 OC-48/STM-16 or 64 OC-192/STM-64 interfaces as well as OC-12/STM-4 and OC-3/STM-1 optical interfaces for legacy infrastructures. All of the CoreDirector's optical interfaces can be configured as concatenated for wavelength switching, or channelized for grooming and switching at STS-1 granularity. CoreDirector therefore incorporates the functionality of SONET/SDH add-drop multiplexers, digital cross-connect systems, and optical cross-connect systems.

CoreDirector's protection options support a variety of differentiated service levels based on service priorities and required protection levels. The system supports a wide range of protection schemes. Additionally, CoreDirector's automatic grooming takes optimal advantage of network bandwidth and maximizes bandwidth availability.

Cisco

Cisco has aggressively made its presence known in the SONET and SDH marketplace through both the repositioning of its highly diverse products and the acquisitions of a number of strategically valuable companies designed to round out the product line. The company's 15XXX product line is designed to satisfy the requirements of both metro and long-haul providers. The Cisco ONS 15540 Extended Services Platform, for example, is a modular, scalable next-generation DWDM platform that integrates data networking, storage, and information streaming over a high-bandwidth intelligent optical infrastructure that supports both packet and wavelength transport. On the other hand, the ONS 15454 SONET multiplexer is an evolutionary optical transport platform that enables networks to carry data, voice, and video traffic. The ONS 15327 is a metro edge optical transport platform, equipping SONET/SDH networks with integrated optical networking and multiservice capability.

Additionally, Cisco's product line includes a broad suite of ancillary and support devices that enable customers to build complete network solutions across the entire range of transport regions.

Cyras Systems

Cyras Systems is a remarkable company that has already demonstrated a system that can scale to OC-768/STM-256—a 40-Gbps system.

The MetroDirector K2 Next-Generation SONET Platform is a data-optimized SONET product with DWDM capability that integrates all the functions provided by stand-alone digital cross-connect systems, SONET *add-drop multiplexers* (ADMs), multiservice data switches, and DWDM terminals.

For data applications, The K2 offers scalable interfaces from DS-1 to multiple OC-12c ATM interfaces. TDM services are supported at rates from DS-1 to multiple OC-192s. Per-shelf port densities are available from 336 DS-1s to multiple OC-192s.

Separate SONET channels are available for TDM, ATM, IP, and LAN protocols. Using concatenated channels, bandwidth can be allocated on an as-needed basis. This bandwidth optimization technique offers granularity down to the STS-1 level, and enables extra traffic to be transported across SONET protection channels.

Fujitsu

Fujitsu's FLM series offers a wide array of add-drop multiplexers for all manners of installation including the FLM 150, the FLM 600, and the FLM 2400. Supported architectures include terminal, unidirectional path-switched ring, linear add/drop, optical hub, and in-service upgrades between architectures. The devices also support a wide array of tributary interfaces including DS1, DS3, EC-1, OC-3, Ethernet, and DS3 UNI for ATM.

Geyser Networks

Geyser Networks' Optical Services Manager™ (OSM 4800) is a multiservice platform that combines DWDM, a SONET cross-connect system, a SONET ADM, an MPLS router, and an *Integrated Access Device* (IAD) all in a single chassis. With the OSM 4800, service providers can build SONET rings to transport a wide array of services including TDM, ATM, frame relay, Packet-Over-SONET, Transparent LAN, streaming audio and video over IP, as well as SLA-driven IP/Ethernet services and MPEG-II video services. An integrated 16 wavelength DWDM system combines multiple wavelengths to form multiple logical rings that increase the OSM 4800's bandwidth capacity.

Hitachi

The Hitachi AMN 5192 Advanced Multiservice Node is designed to address a wide array of applications from long-haul transmission to metro. It combines OC-48, OC-12, and OC-3 multiplexing with an array of advanced capabilities.

Lucent Technologies

Lucent offers a wide array of SONET and SDH products, many of them marketed under the WaveStar product line. Lucent's products span the full range of SONET and SDH capabilities, with a strong emphasis on emerging capabilities such as 40-Gbps systems.

Mayan Networks

Mayan has chosen to recognize the viability of existing protocol structures and incorporates IP, ATM, and frame relay interfaces in its product. Mayan's Unifier product is designed to aggregate customer traffic using IP protocol. The Unifier SMX is unique in its capability to aggregate, groom, switch, and provision voice and data traffic over metropolitan access and transport networks.

The device optimizes both new and existing SONET/SDH ring or point-to-point topologies and enables carriers to evolve to next-generation metro technologies such as IP over DWDM as demand arises.

NEC

The SMS-150V can multiplex a variety of tributary types including 2-Mbps and 34/45-Mbps payloads, which enable the device to adapt to changing traffic patterns.

The SMS-150V has standard STM-1 optical interfaces with VC12/3 path protection and supports metro and access networks, multiple tributary signals (63 × 2 Mbps, 3 × 34 Mbps, 3 × 45 Mbps), and reliable network protection schemes.

Another product, the SMS-600V, is an STM-1 or a STM-4 multiplexer. It supports terminal functions, add-drop, ring, local cross-connect capability, and regenerator functions.

This product combines the capabilities of an STM-1 add-drop multiplexer with an STM-4 add-drop multiplexer. The SMS-600V multiplexes a variety of tributary types, including 2 Mbps, 34 Mbps, 45 Mbps, and 139 Mbps.

The next product in the line, the SMS-2500A, is an STM-16 multiplexer for backbone networks. It offers tributary interfaces that include 139 Mbps, STM-1e, STM-1o, and STM-4. Cross-connect capability is also included.

Nortel

Nortel has an extensive line of both SONET and SDH products. The company currently enjoys the number one position with its 10-Gbps products, holding a whopping 95 percent of the marketplace. Products include both long-haul and metro multiplexers.

Redback/Siara

The Redback® SmartEdge™ 800 is a multiservice optical networking platform that helps service providers improve the economics of their SONET/SDH networks while providing a migration path to IP-based services. The SmartEdge 800 delivers multiring management for SONET and SDH networks. The system uses a TDM and packet-processing architecture that delivers maximum system resiliency.

Service providers can use the SmartEdge 800 to build a data-capable metropolitan optical network. The service provider can start by increasing the density and flexibility of its core metro rings using OC-48/STM-16 or DWDM. The service provider could then build out its metro ring infrastructure. At the edge of the network, the provider can deliver multiple access rings per shelf to customer premises. The entire network can thus be managed through a single interface.

Resilient Packet Ring Alliance

One area of activity that should be mentioned in our discussions of effective bandwidth management in SONET and SDH networks is the effort currently underway by an industry organization known as the *Resilient Packet Ring* (RPS) Alliance. Based on work outlined in IEEE standard 802.17, the Resilient Packet Ring Alliance is working to create a network solution for moving data traffic across a shared network infrastructure with recovery times equivalent to those available in SONET and SDH networks. Instead of optimizing networks for the transport of traditional TDM traffic, the RPR strategy is to design a network that is optimized for carrying packet-based services that provide dynamic bandwidth provisioning; that utilizes auto-discovery at ring startup; that employs a ring protection protocol to ensure survivability; and that is designed in such a way that little capacity engineering is required. RPR is a Layer 2 technique designed for use in metro applications, offering diverse QoS, carrier class protection, and less expensive network operations. The technique relies on *Spatial Reuse Protocol*, a technique that enables bandwidth to be reused on a ring. In spatial reuse environments, the destination node employs a technique called destination stripping, in which the message is removed from the ring upon arrival rather than being allowed to continue on to the originating node to signify its correct arrival at the destination.

The RPR Alliance, which can be found online at **www.rpralliance. org**, includes such members as Cisco, Nortel, Avaya, Cyras, Riverstone, Alidian, and a number of others.

Sycamore

Sycamore is primarily targeted at the wide area environment. Both Storm Telecommunications and 360Networks have installed Sycamore's SN 16000 switches at the heart of their mesh networks. The company's recent acquisition of Sirocco Systems provides Sycamore with a viable entree into the SONET and SDH world; expect to see more from this company.

Ultra Fast Optical Systems

This company has an intriguing technological proposition. Many industry pundits believe that SONET will reach the limits of its transport lifetime at or around 40 Gbps. In response, Ultra Fast Optical Systems has announced *Optical Time-Division Multiplexing* (OTDM), a technology that is similar to SONET/SDH in that it has the capability to transport low-speed services in high-speed channels. However, it differs from SONET/SDH in that it is an all-optical solution, which enables it to operate at extremely high speeds. According to a paper presented at the European Council on Optical Networking (ECOC 2000), OTDM can reach speeds as high as 640 Gbps—and that's one wavelength! OTDM interworks seamlessly with DWDM to deliver terabits of bandwidth.

Of course, other vendors are in this game; we have attempted to show a sampling of them to illustrate the diversity of the products that are available.

In Summary

SONET and SDH are a set of standards that define a high-speed, synchronous network and a range of capabilities based on the robustness of optical fiber. The standards address a unique payload-mapping scheme, a well-defined set of Operations, Administration, Maintenance, and Provisioning messages, and a standardized hierarchy of data transmission rates.

If you're familiar at all with the development of the telephone network, then you know that digital carrier systems such as T1 and E1 were first deployed as intra-network pipes for high-volume trunking. Over time, they leaked out of the network and, like the tentacles of a giant squid, extended themselves to the customer. SONET and SDH are following a similar path. Initially, they have found applications inside the telco networks as a trunking medium, but are slowly migrating out to the customer as end user applications emerge and SONET/SDH-compliant equipment becomes available.

In today's networks, the use of scarce and expensive transmission facilities is optimized by concentrating multiple streams of data or voice conversations on a single channel—a process called multiplexing. In digital carrier environments, a variety of incoming signals are combined by a multiplexer into a composite signal that is transmitted over a T1 or E1 facility. This enables multiple voice and data conversations to be carried simultaneously by a carrier, reducing the cost to the service provider, and of course, to the customer.

This composite signal consists of multiple 8-bit samples from each of the channels of the multiplexer, plus framing information that marks the beginning of the frame of data. At the central office, these multiplexed signals are further combined to create higher bit rate transmissions.

In countries that rely on the T-Carrier hierarchy, the devices that perform this *cascaded multiplexing* function are called M13 multiplexers because they combine four DS1 signals to create a DS-2, then seven DS-2s to create the aggregate DS-3.

The resulting 44.736-Mbps signal is transported to its destination, where the receiving central office equipment, typically another M13 mux, disassembles the signal and routes the various subcomponents onto their final destinations.

Needless to say, telecommunications networks are complex and highly interwoven creatures. The components of a multiplexed bitstream can originate from a variety of networks, and as a result, may have slight timing differences among the signal components. To ensure the timing integrity of the network, incoming bit streams have to be rate-aligned through a process called bit stuffing. As the name implies, bit stuffing procedures actually stuff bits into the real data stream to ensure that the different components are properly aligned and "dancing to the same network tune." The resulting alphabet soup of data bits and stuff bits is then transmitted over the network.

The genesis of a high-speed DS-3 frame from intermediate DS2s, which in turn, originated from individual DS1s, is quite complex. By the time the carrier system has created the final DS3 signal, the location of each composite source is buried under three separate layers of overhead and three independent framing systems. In order for a customer to access their information in this technological alphabet soup, the entire three-stage process must be reversed. This requires equipment dedicated solely to the task of demultiplexing and remultiplexing the stream—a pair of back-to-back M13 multiplexers.

The difficulty caused by this multilayered signal building process becomes obvious every time components of a signal need to be dropped out. At these add-drop points, the entire complex signal must be decomposed, the various control and stuff bits identified and backed out, the component in question identified, selected, and dropped out, and the remaining pieces reassembled and sent on their way. This process, called *back-to-back multiplexing*, is unwieldy, hardware intensive, time consuming, and expensive. SONET and SDH eliminate the need for this process. They represent a move forward along the evolutionary path, and are designed to accommodate all network standards, easily and transparently.

Other factors have driven the remarkably rapid development of the SONET and SDH standards as well.

Advancements in transmission technologies and the development of broadband services have increased customer demand for bandwidth, and optical transmission is clearly a cost-effective way to provide it. Both fiber and the terminating electronics for optical networking have experienced remarkable advances, feeding the evolution toward fiber-based systems and all-optical networking.

Another factor that has positively affected the successful deployment of SONET and SDH is the proliferation of highly capable, low-cost chipsets for high-speed networking applications. Like any new technology, the creation of mass-produced *very large-scale integration* (VLSI) technology has caused the price of electronics to fall precipitously. Advances in the technologies associated with producing the fiber itself have seen its cost fall from a level that was once prohibitive to the point that the cost of a deployed fiber mile is comparable to copper, perhaps cheaper in some cases. On a bandwidth-delivered basis, fiber is obviously *far* less expensive than copper.

Another driver is the evolving role of fiber itself. In keeping with current and planned advancements in optical technology, it is a firmly held belief that optical fiber will be the medium of choice for interconnecting high-speed, high-volume central office switches, corporate facilities, perhaps even customer access devices. True photonic switches that switch light pulses without first converting them to electrical signals are commonplace today, and optical routers are not far behind; a fiber backbone arrangement to interconnect them is clearly necessary if their full potential is to be realized.

Finally, as networks become faster, more diverse, and more interconnected, some form of centralized and universal network management system begins to look rather attractive, particularly given the overwhelming focus on QoS as the principal deliverable. In fact, SONET and SDH provide the underpinnings for a network architecture that satisfies all of these requirements.

Two key factors continue to play leading roles in the accelerated rollout of fiber systems and the demand for SONET and SDH.

The first is the never-ending entrepreneurial development of high-bandwidth applications. Among these are high-speed LAN interconnect for the metro market, graphic and video-intensive multimedia applications, interactive video, television, movies on demand, and a host of others.

As applications emerge, the ability to provision bandwidth on demand for these applications becomes a major market driver, and corporations have made their desires for immediate, high-quality transport known rather loudly.

Another driver behind the remarkable success of SONET and SDH is expense. As the amount of deployed fiber has increased, the cost of bandwidth has plummeted. Bandwidth, then, becomes a commodity. The economic constants of supply and demand are well known here: the more available a commodity becomes, the less expensive it tends to be. Bandwidth certainly follows this model.

Of course, technology, falling prices, and applications development haven't been the only contributors to SONET's rapid evolution. As with many innovations, the strength of the marketplace has played a key role. In 1984, shortly after the divestiture of AT&T, MCI's Bill McGowan went before the *Interexchange Carrier Compatibility Forum* (ICCF) to ask for their assistance on a growing problem. The government's Equal Access rulings ensured that all interexchange carriers would have equal access to each local exchange carrier's customers, thereby guaranteeing fair market access to both AT&T and all other long-distance carriers, MCI among them.

The problem was that, although the rulings guaranteed points-of-presence in the local exchange central offices, they left the onus of interconnection, that is, equipment compatibility, up to the interexchange carrier. As you might imagine, this became rather expensive because quite a variety of deployed equipment by that time would not interoperate. Every POP that MCI established meant that they had to purchase the right termination equipment in order to interconnect with the local exchange carrier.

To resolve this expensive and complex (and clearly dead-end) issue, McGowan called for the creation of a mid-span meet standard that all network equipment manufacturers could design interfaces for, thus allowing for vendor-independent interconnection of fiber transmission systems.

The idea was a good one whose time had come. Bellcore, the *American National Standards Institute* (ANSI), and the *International Telecommunication Union* (ITU) quickly formed study groups to research the concept. After a certain amount of political puffery and positioning, the North American standards bodies and the ITU were able to define a transmission hierarchy that became the international standards known as SONET and SDH.

These standards were initially created as two distinct phases. The first dealt with physical and hardware parameters such as signal construction, multiplexing, optical concerns such as laser power and pulse shape, and payload mappings. The second phase was designed to refine the physical parameters and clearly define the protocols and messages that would be used to transport network management and error control messages.

This second phase turned out to be more ambitious than they originally thought, and in due course, it was divided into two pieces. Phase two refined the specifications for the electrical/optical interface that would play a major role in SONET/SDH and identified the necessary Operations, Administration, Maintenance, and Provisioning messages that would be required in this new network. Phase three, released in early 1992, defined more specific OAM&P message sets, and outlined the use of SONET's data communications channels.

In 1988, the initial standards for the Synchronous Optical Network were released, and by 1992, the standards were finalized. This incredibly swift development cycle is a testimony to the need that existed for such a set of capabilities.

In current systems, multiplexing plays a key role in enabling service providers to make the best possible use of scarce and expensive transmission facilities. The multiplexing process is complex, however, and uses a mixture of data and control information to keep track of individual signal components. The back-to-back multiplexers that are capable of unscrambling the composite signal are necessary, but expensive. Network engineers recognized the need for a better way.

Other factors influenced the eventual birth of SONET as well. These included the development of broadband services, such as medical imaging, increasing customer demand for bandwidth to accommodate those applications, rapid advances in fiber technology, VLSI, and optoelectronics, and an attendant drop in cost, a very real need for standard transmission rates above DS-3, and a growing need for centralized network management in complex systems.

All of these paved the way to the development of SONET, but the effort became real in 1984, when MCI asked the ICCF to push for the development of mid-span meet capability, which would eliminate the problems encountered in multivendor environments. The ICCF and other standards bodies agreed, and the SONET standards were rolled out in three evolutionary phases.

Let's look at the standards themselves. Mid-span meet capability overcame the fact that most network equipment relied on proprietary transmission schemes. This meant that interoperability could only be achieved through the very costly expedient of matching equipment on a like-for-like basis. Clearly, the standard made good economic sense for the carriers, but mid-span meet was only one of the challenges that SONET was designed to overcome.

The many components and sub-networks of today's transmission systems have sophisticated network management systems, which, on the one hand, is a good thing. On the other hand, the number of management systems tends to grow in proportion to the complexity of the network, making the network manager's job something of a nightmare. The typical Network Control Center looks like a television display in a department store, with messages and information displayed in every possible format imaginable. It's up to the network manager to assemble and collate all the incoming messages, then make decisions based on their content. There has to be a better way.

We've already discussed back-to-back multiplexing, the cumbersome technique used in today's networks to add or drop signal components. In many cases, as applications emerge, customers become bandwidth limited. Major network retrofits and workarounds are often needed to comply with customer demand as bandwidth-hungry applications become more and more common.

In the post-SONET and SDH world, true vendor independence is finally a reality.

Additionally, SONET provides a standard suite of maintenance and management messages. This means that network management and provisioning can be done from a single system on a network-wide basis. Incoming status messages from cross-connect systems, fiber systems, multiplexers, and miscellaneous information from far-flung central offices can now be displayed on a single monitoring device in a standard format, thus consolidating the information into a usable structure. This enables network management personnel to react to network difficulties on a global basis, instead of on a component-by-component basis—sort of a "Zen and the Art of Network Management" concept.

SONET and SDH's sophisticated multiplexing techniques enable signal components to be added and dropped from the bitstream, but it does so without back-to-back multiplexers. The fiber-based transmission hierarchy described in the SONET and SDH standards provides customers with virtually unlimited bandwidth—and more can be added routinely.

The key to SONET's and SDH's acceptance as international standards lies with its logically structured hierarchy of transmission rates. Standards designers knew that the system they created had to accommodate both North American and European transmission schemes, with rates ranging from DS1, to DS3, to Europe's 139.284 Mbps CEPT-3 signal—and beyond.

To transport these disparate signals, SONET relies on a 51.84-Mbps basic building block, called *Optical Carrier Level One* (OC-1). Multiple OC-1s can be combined to create higher rate signals. OC-3, for example, is exactly three OC-1s. An OC-3 signal easily accommodates a CEPT payload —as easily as an OC-1 transports a DS3 signal.

Something of a disparity appears between the OC rate and the signal that it transports. OC-1 is 51.84 Mbps, whereas DS3 is only 44.736 Mbps. A similar difference exists between OC-3 and the CEPT-3 signal. SONET has a quite sophisticated yet simple way of placing payload into a frame structure that doesn't actually care where the data starts. In fact, the payload can move around within the frame, and SONET will keep track of it, thanks to the payload pointer.

In SONET and SDH, some of the bandwidth is overhead—a lot of it, in fact. In a 51.84-Mbps OC-1 signal, for example, over 2 Mbps is overhead. This stems from the fact that SONET and SDH are based on a philosophy that bandwidth is cheap—and it's a good thing because they use a lot of it for overhead. Among other things, this overhead is what SONET and SDH use to keep track of the payload.

In traditional networks, systems rely on the hardware's capability to discern certain patterns in the data that denote beginnings and ends of data components. T1 systems, for example, and others like them, rely on a repeating pattern of framing bits to ensure frame integrity. By counting data bits between framing bits, the network knows where the data begins and ends.

SONET and SDH, on the other hand, steal an idea from the world of software: the pointer. The upshot of this is that the data can effectively start anywhere within the frame, and the pointer will always indicate the first byte of the payload. If the data shifts slightly, due to a shift in phase, no problem—the pointer adjusts as well. In this way, SONET and SDH deal effectively with minor phase discrepancies. If a disparity is detected, the pointer moves forward or backward, thus speeding up or slowing down the payload.

Today's network components are extremely stable and rely on accurate internal clocks for timing consistency. As signals pass from device to network device, they are often incrementally affected by minute variations in the power supplies within each device. This causes the signals to shift slightly in phase, a condition that in traditional networks can cause serious problems.

SONET and SDH rely on a timing scheme known as *plesiochronous timing*. In a plesiochronous environment, the network knows that slight timing discrepancies will exist between the components. Instead of viewing this as an error condition, a plesiochronous network actually enables the various signal components to float slightly relative to one another. In SONET and SDH, the payload pointers manage this function. Thus, fluctuations that at one time caused serious timing slips, today are completely transparent.

We've spent a lot of time talking about SONET and SDH's abilities to transport super-rate services. What about sub-rate services, such as T-1 or E-1? In fact, SONET and SDH transport T1 and other lower-rate signals quite nicely, using a mechanism called *virtual tributaries* (VTs) or virtual containers.

Virtual tributaries/containers are like small buckets that carry small payloads. When a SONET system, for example, is transporting VTs, the 90-column STS is subdivided into seven 12-column groups called VT Groups. This consumes 84 columns; of the remaining six, three are Transport Overhead, one is Path Overhead, and the other two are called *fixed stuff*. They are not needed and are ignored by the SONET system.

Let's talk a little more about that structure. Four types of virtual tributaries are available: VT1.5, VT2, VT3, and VT6. A single, 12-column VT Group can accommodate four VT1.5s, three VT2s, two VT3s, or one VT6.

Consider, for example, a virtual tributary group that is transporting VT1.5s. Within each VT1.5, a DS1 signal can be mapped. The standards do a good job of illustrating the different mapping techniques that can be used for different payload types. For example, they carefully define both bit and byte-synchronous mappings for DS1, as well as asynchronous DS1 mappings and mappings for higher rate services, as well.

SONET can operate in either of two modes when transporting virtual tributaries. In locked mode, payload modules are mapped directly into STSs, thus facilitating the network's capability to distinguish them and reduce overhead. In fact, the goal is to eventually reach a point where network-timing sources are stable enough to eliminate the need for the floating pointer completely, and operate solely in locked mode. If the payload doesn't drift, then obviously a floating pointer is not needed.

In floating mode, virtual tributaries are mapped into STS payload envelopes on an "as they arrive" basis, meaning that components will vary slightly from one another in phase. Virtual tributary payload pointers then help to keep track of the VTs themselves. Floating mode also eliminates the need for hardware buffering because one key reason for buffering is to ensure that incoming data streams are frame aligned before being placed out on the network. Similar rules apply for the SDH world.

Finally, we saw that SONET and SDH have begun to dominate two key regions of the network: the metro edge and the long-haul core. Both areas are important, and as a result, we have seen the emergence of two principal network architecture models: the ring and the mesh. Rings are primarily finding a home in the metro market, whereas meshes serve the long-haul market well. Vendors have lined up behind both topologies and are rolling out innovative products quickly.

So, what is the end result of all this? It is clear that both SONET and SDH will enjoy a place in the network technology court for some time to come, although their shape, size, influence, and roles may change somewhat. They represent the bulk of the embedded network that is currently in place, and as such are extremely capable at what they do. The responsibilities for which they were created are still very much in evidence, particularly because circuit-switched voice still represents the bulk of all revenue-generating traffic carried across modern telecommunications networks. However, as data in its many forms—including IP voice and multimedia—continues to evolve and demand a bigger piece of the transport pie and generate a larger, more noticeable chunk of network revenues, both SONET and SDH will be called upon to evolve in lockstep. Network providers, too, will have to evolve, and that promises to be the greatest challenge of all. They have enormous investments in a network architecture that has survived the test of time for nearly 30 years, and the changes required to modify the network for new services will not be easy because it will not only include technological changes, but managerial, functional, procedural, and human resources modifications as well. The changes will happen; there is no question about that. The issue at this point is how quickly the changes will occur, how long the market will wait for the changes, and how well the hardware vendors will step up to the challenge of providing next-generation SONET and SDH hardware that will seamlessly adapt to the demands of the marketplace.

In our final chapter, we examine SONET and SDH applications, the *real* value behind the technology.

SONET
and SDH
Applications

In preceding chapters, we have examined the market drivers that are helping to make SONET and SDH critical technologies in the modern telecommunications marketplace; the underlying technological details that make them work; and the companies that manufacture products and use those products to create, manage, and maintain the modern global telecommunications network infrastructure. In this chapter, we turn our attention to the applications that take advantage of the network.

If we go back and revisit the chapter in which we discussed the drivers behind the success of both SONET and SDH, we find three common themes among all of them:

- To meet the growing demand for bandwidth
- To reduce the overall, aggregate cost for bit transport
- To support the transport of differentiated services

We will examine each in turn.

Differentiated Services

Meet the growing demand for bandwidth: This stems from the inexorable growth in network-dependent applications that has been an ongoing phenomenon for more than 10 years. The global network has passed through a number of availability crises stemming from bandwidth demand outpacing bandwidth availability, but today that is less of a concern because of (1) the availability of in-place fiber, and (2) the capability to provision multiple channels over each fiber through the judicious deployment of DWDM. The answer then is to build higher capacity TDM systems, manufacture DWDM systems with higher channel counts, and ultimately enjoy systems with higher lambda counts per fiber, and higher bit rates per lambda. When this becomes a reality, the challenge of meeting demand for bandwidth will no longer be a problem.

Reduced cost-per-transported bit: The costs associated with building, maintaining, and operating network transport systems derive from several sources, including human resources, capital expenditures for hardware and software, facility construction and management, and ongoing operational expenses. Three steps can be taken to reduce these costs relatively quickly: reduce operations costs, which involves properly managing in-place network resources to ensure that they are being properly maintained, provisioned, administered, and billed; reduce floor space requirements, which can be inordinately costly, especially for service providers that do not have their own central office rack space and must therefore rent it from an incumbent provider; and, finally, reduced equipment costs, which can be achieved through careful network planning, proper maintenance, and discriminatory equipment selection. Proper hardware selection, for example, can result in the construction of a network that is relatively future-proof and that will avoid the need for "forklift component upgrades" in the short to medium term.

Support of differentiated services: A clear differentiation has been recognized among the traffic types carried across most networks. These include circuit-based services such as voice traffic and others that require traditional TDM transport, packet-based services to serve the requirements of emerging data applications, particularly IP, and wavelength-based services, such as those that require inordinately high bandwidth and therefore rely on the capabilities of multichannel DWDM transport. It should be noted that within each of these service types, an individual *quality of service* (QoS) granularity can be achieved, thus guaranteeing the ability to provide transport service to a broad array of traffic types.

Within the transport domain of most network service providers, a subset of service types has emerged that can be easily differentiated by application type. These include circuit-, packet-, and wavelength-based services, described earlier; services that address the emerging requirements of the metropolitan network environment, including Ethernet-to-Ethernet connectivity; the services that reside in the core of the optical network; and the relatively new *optical area network* (OAN) services arena that supports a wide array of applications including *storage area networks* (SAN), Web hosting, bandwidth trading, video applications, and data center interconnection. We will examine each of these with an eye on the applications that each makes possible.

Circuit-Based Services

Circuit-based services are characterized in the following fashion. They typically transport traffic from one physical port on a SONET or SDH device to another, and in general, circuit-based services support the following interface types, as shown in Figure 7-1: DS-3, EC-1, OC-3, OC-12, OC-48, OC-192, Gigabit Ethernet, STM-1, STM-16, and STM-64. In most cases,

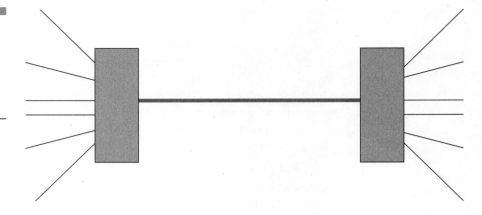

emulating the private line services that they replace, circuit-based services support transmission rates ranging from 45 Mbps to 1 Gbps, usually in 45-Mbps increments. These services generally transport client bandwidth transparently from a source point to a destination point, provide fixed delay from end-to-end, and offer measurable and predictable QoS based largely on protection scheme capabilities and some measurable maximum bit error rate.

Implementation Options

Circuit-based services can be implemented in a variety of ways. Traditionally, they are provisioned over standard time-division multiplexed facilities such as SONET and SDH. More and more, however, the marketplace is asking for alternatives to these legacy solutions, such as ATM and IP. One of the services that ATM is capable of providing, shown in Figure 7-2, discussed earlier in the ATM chapter, is called *circuit emulation service* (CES). With CES, ATM cells are filled with TDM traffic for transport across the ATM *wide area network* (WAN). Bandwidth is moderated by changing the number of cells per second that are transmitted and the number of bytes of user payload that are placed into each cell. By controlling these two parameters, the overall throughput of the network can be carefully controlled.

Another potential solution is to use IP packets as the transport mechanism. IP is a layer-three solution that has garnered a great deal of attention

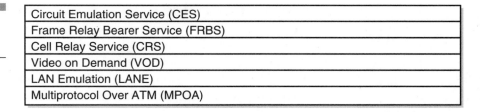

Figure 7-2
The many services
provided by ATM.

Circuit Emulation Service (CES)
Frame Relay Bearer Service (FRBS)
Cell Relay Service (CRS)
Video on Demand (VOD)
LAN Emulation (LANE)
Multiprotocol Over ATM (MPOA)

in the last few years because of its integral position in the Internet. It is a connectionless solution, however, which means that the capability to guarantee latency and QoS is severely limited, and virtually impossible (today) if implemented across the Internet properly. The only viable solution is to use proprietary options such as Cisco's Tag Switching concept that is extremely effective but indeed proprietary. Cisco has worked closely with other companies in the industry to develop a non-proprietary solution called *Multiprotocol Label Switching* (MPLS) that accomplishes the same task in an open environment, shown in Figure 7-3. Nevertheless, delivering guaranteed bandwidth and QoS over an all-IP network remains a task that most network providers consider Herculean.

Ultimately, there is a single truth that cannot be ignored: neither IP nor ATM scale well at bandwidth levels in excess of OC-3/STM-1 (155.52 Mbps). TDM remains the best solution for the delivery of legacy traffic.

Figure 7-3
The evolution
to MPLS.

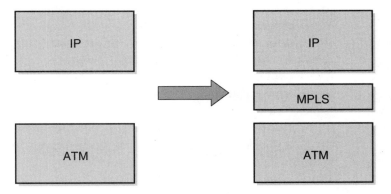

Measuring Service Differentiation

Circuit-based services, like all telecom transport services, require a set of metrics that provides network planners and administrators with the wherewithal to measure the relative effectiveness of deployed technologies. These measures are most effective when they align with and provide a measure against the market drivers discussed earlier: support for the delivery of differentiated services, reduction of the overall aggregate cost to deliver bits from a source to a destination, and a successful capability to meet the never-ending demand for more and more bandwidth.

The first of these, support for a diversity of service types, can be measured in several ways. These include a high degree of flexibility with regard to bandwidth management; support for a diverse array of service interfaces, including DS-3, OC-1, OC-3, and above; and the capability to provision extremely flexible and granular bandwidth with as close to an "on-demand" service level as possible, including both channelized and concatenated services. Support for many service types can also be measured by the capability to assign variable bandwidth levels to service rate-independent ports, including such interfaces as the various flavors of Ethernet, and the capability to offer a variety of protection schemes, including BLSR, UPSR, 1+1 for mesh networks, 1:N restoration, and 0:1 unprotected facilities. Customers are looking for diverse service types and demand has arisen for the capability to pay at different billing levels for different protection levels.

The task of reducing the overall cost per transported bit can be accomplished in a variety of ways. These include the capability to provide facile scalability with a pay-as-you-grow option for service providers and the implementation and use of well-designed management systems as a way of performing end-to-end management and well-timed scalability as cost-control measures. Reducing the overall cost per transported bit can also be done through hardware designs that increase the port density of deployed devices, thus reducing the floorspace footprint required to deliver scalable bandwidth. It can also be accomplished through the general use of modern switch, multiplexer, and cross-connect designs as a way to future-proof the network and ensure that network providers are enjoying the most bang for their bandwidth buck while offering multi-vendor interoperability where possible.

Deployed Applications

Among the many applications that have been posited for use in the circuit-based environment, several have emerged with great potential. These include the backbone feeder application, the submarine long-haul application, and the metro core. We will examine each in turn.

Backbone Feeder

In a typical backbone feeder implementation, traffic is delivered by low-speed rings or linear spurs and is aggregated into a high-bandwidth transport service such as the 10-Gbps backbone ring shown in Figure 7-4. In this illustration, a 10-Gbps backbone installation is interconnected to the low-speed subtending rings via a two- or four-fiber BLSR. Multichannel DWDM devices create additional bandwidth on the ring, and optical cross-connect devices provide the mechanism to effectively deploy the additional

Figure 7-4
Traffic aggregation for backbone transport.

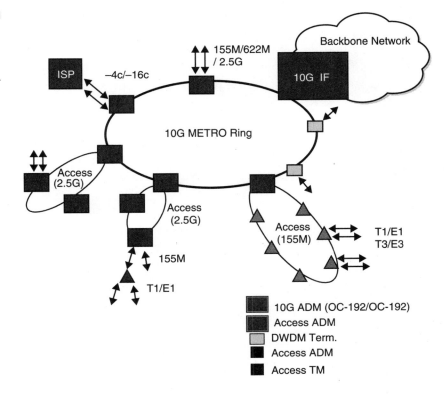

bandwidth. The low-speed rings feed traffic into the backbone via the 10-Gbps ring, as do the other interfaces shown—EC-1, DS-3, OC-3, OC-12, OC-48, and OC-192—all with diverse protection levels as requested by the customer. By offering compatible optical interconnection between the various devices in this network, customers can reduce the cost of network deployment through the elimination of such devices as optical translators and other adaptation devices. Finally, element and network management systems add an additional degree of control for service providers.

A word about the ongoing saga of the BLSR versus the UPSR: Bidirectional line switched rings are often used in interoffice applications to move traffic between nodes or central offices where it is more cost-effective and strategically positioned than a UPSR ring. This is because they split the transported traffic between the two rings as a way of guaranteeing route diversity for critical traffic. As we said earlier, two versions of BLSR are available: the two-fiber implementation and the four-fiber ring. The two-fiber implementation has a lower startup cost but a lower overall total capacity. It has the capability to provide protection against single points of failure. The four-fiber ring, which provides significantly greater capacity than the two-fiber ring, has a higher startup price. The four-fiber ring is often favored by long-haul carriers because maintenance can be performed on one span without affecting the rest of the ring.

Transoceanic Applications and the Transoceanic Protocol (TOP)

Transoceanic applications present a variety of issues that are frankly fascinating and challenging—in fact, vexing is probably not too strong a word to add to the list. Let's begin with a brief discussion of the history of submarine optical networks.

The first submerged copper cable was installed by Cable & Wireless in 1856 between Newfoundland and Ireland to provide teletype service between the two continents. It was finally decommissioned in the mid-1920s (although according to my sources it still works today!). It provided superb service and served as a good indication for what was to come.

In 1943, the UK installed the first undersea cable with integral repeaters between Port Erin and Holyhead. In North America, the first submarine cable was installed in 1950 between Havana, Cuba and Key West, while the first transatlantic cable (TAT-1), which was coaxial, went live in 1956 between the UK and Canada. It had capacity for 36 simultaneous phone circuits (all of which were rarely used simultaneously) and had repeaters installed roughly every 50 miles. TAT-7, which went live in 1983 and is still in use, uses repeaters every 5 miles and was the last coaxial cable put into service.

Before long, optical networking caught the attention of Bell Laboratories as an effective voice and data transport solution. By the mid-1970s, they were working to perfect optical transmission for use in TAT-8, the first transoceanic cable planned to be based on optical fiber. The design called for three pairs of single-mode fiber operating in the 1300 nm window, with two live pairs and a third serving as a spare. The cable was designed to transport voice and data at 278 Mbps on each pair, delivering an aggregate data rate of approximately 550 Mbps. With voice compression technology, the cable permitted the simultaneous delivery of 40,000 voice calls—a far cry from TAT-1's 36. TAT-8 required repeaters every 30 miles and demanded very respectable mean-time-between-failure measurements that guaranteed a submerged lifespan of 25 years and the need for no more than two recoveries of the cable during that time for repairs. The cable became operational in 1988, and with its success, a raft of new applications and services emerged designed to take advantage of this almost incomprehensible volume of bandwidth that had finally become available.

In the early 1990s, a third transport window became available at 1550 nm, making it possible to use EDFA technology and increase the overall data rates of installed cables. TAT-9 and TAT-10 offered 565 Mbps of bandwidth and repeater spacing well in excess of 60 miles. TAT-12 and TAT-13, a remarkable optical loop design that connects the U.S., the U.K., and France, added EDFA technology and offered 10 Gbps of overall bandwidth.

More recent cables provide optical add-drop multiplexing and even greater bandwidth, such as 640 Gbps in the case of Project Oxygen. Oxygen is an ambitious $14-billion project funded by venture capital and intended to interconnect the world with flat-rate, distance-insensitive transport. Lucent Technologies and Corning have been selected as major suppliers of switching equipment and non-zero dispersion-shifted fiber. Although the project is behind its original schedule, it is moving forward, and project managers are optimistic that they will complete the project as originally thought. The first leg is expected to be complete in the first half of 2001.

Like *Fiber Link Arouind the Globe* (FLAG), which interconnects Japan with the UK, Project Oxygen's intent is to provide a cost-effective solution to the perceived lack of global bandwidth. However, given the degree to which companies like Qwest, Level3, 360Networks, Tycom, and Global Crossing have deployed bandwidth, that perceived lack may be more imagination than reality. Say's Law, named for French economist Jean-Michel Say, observes that supply creates its own demand, which would naturally lead to the conclusion that any fiber glut is purely momentary.

According to consultants KMI, there will be enough submarine optical capacity installed by 2003 to provide transport for the equivalent of 800 million simultaneous telephone calls. Furthermore, they observe that more than 70 percent additional fiber will be installed in the next 5 years than what has been installed since fiber was first used on the TAT-8 and TPC-3 cable projects in the 1980s. This will equate to approximately $23 billion in installed fiber in the Pacific basin, with more than $50 billion invested by 2003.

Submarine Fiber Installation Techniques

When fiber is to be installed close to shore, it is typically armored and buried in a deep trench to prevent potential damage from ship anchors, dredges, and trawlers. In deeper water, the armor and burial are not required; in fact, it is difficult to work with submarine fiber cables in deep water because the cable is heavy and can break from its own weight while being installed on the seabed.

When fiber is installed close to the shore and is used to interconnect cities along a shoreline, as shown in Figure 7-5, the design is called *festooning*. In a festooned system, the fiber is laid in great loops offshore, usually no more than 250 km long. The loops come ashore at cities where service is required, and amplifiers are usually installed at the landing points as well. In areas where there is concern for the stability of the region and the potential for damage to the cable system is real, a different technique is used. In those cases, optical spurs are installed, as shown in Figure 7-6. These optical spurs do not provide access to the entire fiber; instead, they only bring ashore specific wavelengths selected for use by the region in question. A disruption of the cable at its landing site in this situation would only disrupt the signal going to that particular landfill point, but it would not affect the rest of the cable. In these installations, the trunk (main ring) is usually about 1,000 miles offshore.

Submarine cables reside in one of the harshest and most abusive environments possible. They must withstand high pressure, abrasion, attacks from biting animals such as sharks, unexploded ordinance from past wars, and damage from dredging activities and inadvertently dropped boat anchors.

Figure 7-5
Festooning.

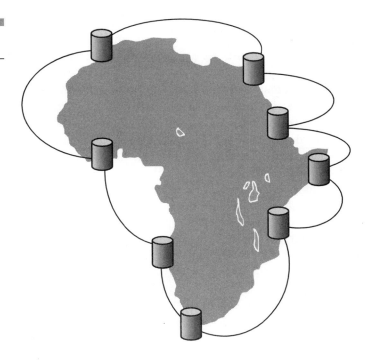

Figure 7-6
Optical spurs.

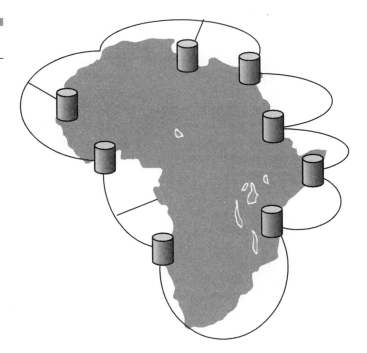

They are also subjected to damage from leakage and earth movement, and must therefore be protected. Figure 7-7 shows the layers that are found in submarine assemblies, including a nylon outer covering, various polypropylene layers, several layers of steel wire armor, a hermetically sealed copper carrier tube, elastic cushioning fibers, a central member known as a king wire, and the fibers themselves.

Long-haul optical spans that crisscross oceans must be carefully designed to take into account geographical considerations, distance, and noise. Long spans must be amplified, but amplifiers amplify noise as well as the desired signal and must therefore be carefully inserted so as to minimize the total noise injected by each of the amplifiers in the span. The noise is cumulative; each added amplifier adds an additional noise component. To reduce the impact of this problem, submarine amplifiers tend to have less gain, allowing for more amplifiers to be used in the span. The gain must be equalized across all wavelengths in systems where DWDM is employed. Network designers must therefore pay close attention to span engineering requirements.

Regeneration must also be done periodically to keep the signal clean; in a typical installation, eight or so amplifiers are installed in series before regeneration is performed, as shown in Figure 7-8. Because DWDM plays a major role in modern optical cable systems, dispersion, four-wave mixing, and other nonlinear effects must be carefully considered.

Figure 7-7
Submarine cable.

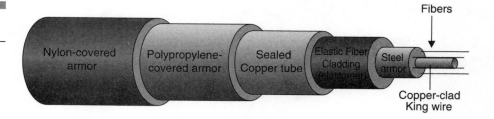

Figure 7-8
Amplification and regeneration in an optical network.

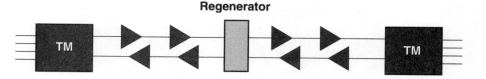

Submarine Challenges

One of the greatest (and perhaps most underestimated) challenges that faces transoceanic cable design personnel is the fact that, in many cases, one end of the cable will be originating SONET traffic while the other will be originating SDH. This presents a unique task for the span termination equipment, which must serve as a gateway between the two largely incompatible protocols. The transoceanic protocol that resides on the transoceanic ring is designed to protect the installation through a very innovative utilization of the SONET/SDH protection bytes. Consider the illustration shown in Figure 7-9.

In this example, we have a four-node dual fiber ring with "landing points" in Burlington (BTV), Boston (BOS), Miami (MIA), and London (LHR). Under normal operating conditions, traffic is carried on the solid outer ring, while the inner dashed ring serves as the backup span. In the event of a span failure between Burlington and London, two scenarios are possible (the nodes have been removed for purposes of clarity). In the first scenario,

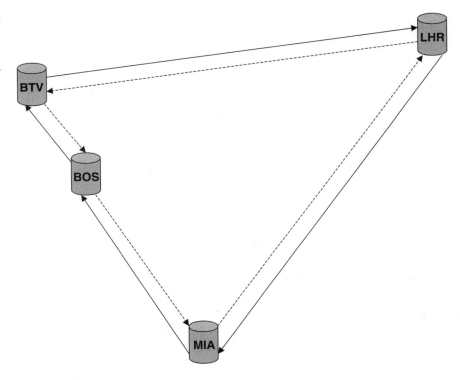

Figure 7-9
UPSR under normal operating conditions.

the Burlington node recognizes the failure on the outbound span and moves to the backup span, as shown in Figure 7-10. This overcomes the problem of the failed span but adds a degree of complexity because now the distance from Burlington to London is significantly longer. Instead of a single hop trip between the two cities, the path now goes from Burlington, to Boston, to Miami, and then to London. In a UPSR configuration, this poses no problem, because the backup span is reserved for potential failures.

What happens, though, if this is a BLSR network? In that case, half of the traffic travels on the primary ring, while the other half travels on the secondary ring. If the network is configured to swap to the secondary span in the event of a failure of the primary span, traffic on the secondary ring could be destroyed.

To avoid this problem, network designers rely on a protocol called the *transoceanic protocol*. Using the transoceanic protocol, channels are assigned on each ring in opposing order. For example, in an OC-48 environment, the primary ring would assign users to channels starting

Figure 7-10
Breach of UPSR.

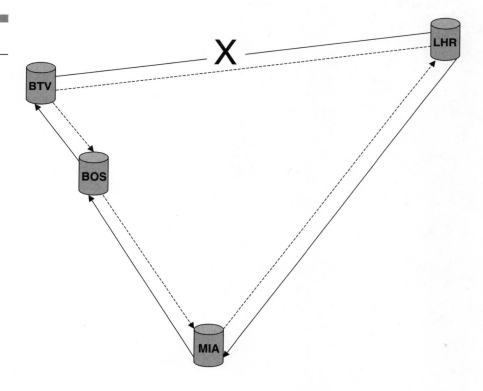

with the highest numbered channel—48 in this case. The secondary ring would be assigned starting with channel 1. The two would then "meet in the middle."

So what is the advantage of this technique? One is that high-priority traffic—that is, traffic with critical QoS requirements—can be placed on the primary ring, while traffic with less critical QoS requirements can be carried on the secondary. In the event that a failure causes the primary to have to switch to the secondary, the high-priority traffic would be 100 percent protected. Depending on the number of channels that the lower priority traffic occupies, it could be unaffected, because the traffic that is switched over will first occupy the channels at the other end of the array of available timeslots, as shown in Figure 7-11.

These implementations usually support a variety of protocols, as they must, given the fact that they straddle SONET and SDH-oriented countries. These protocols include

- STS-1 and AU-3 (HO)

- STS-3c and AU-4

- STS-12c and AU-4-4c

- STS-48c and AU-4-16

- STS-192c and AU-4-64c

- Various Gigabit Ethernet-compatible mappings (STS-3c and AU-4)

Figure 7-11
Channel occupancy
in ring switchover.

Primary Ring Occupied Channels Before Switchover	Secondary Ring Occupied Channels Before Switchover	Secondary Ring Occupied Channels After Switchover	Secondary Ring Traffic Lost Following Switchover
1	1	1	1
2	2	2	2
3	3	3	3
4	4	4	4
5	5	5	5
6	6	6	6
7	-	7	
8	-	8	
9	-	9	
10	-	10	
11	-	11	
12	-	12	

Metropolitan Core Solutions

According to a variety of competitive intelligence reports, as much as 70 percent of all metro and access equipment sales will come from traditional legacy carriers with well-founded, effective, and conservative management systems in place as well as a long-term commitment to either SONET or SDH.

The key challenge that faces metro service providers is the need to efficiently funnel a variety of multi-service traffic types into the carrier's core optical network. Today, services in the metro area are delivered over any number of technologies including TDM, ATM, and IP, all offering a range of bandwidth. The challenge lies in building a network that is not only scalable, but can aggregate that mix of traffic and transport it to the core in an efficient, scalable fashion.

One key observation about optical technology is this: the closer one gets to the end user, the more diverse and complex the network becomes, as shown in Figure 7-12. In the world of optical networking, this is equally true. At the network's edge and within the core, the evolving local loop must support a variety of access technologies and standards including frame relay, ATM, Ethernet, and IP. In response, companies have emerged that focus on the creation of "wide spectrum" network management systems that not only handle multiple protocols and services, but that are flexible, scalable, single-seat systems that reside at the periphery of the network, close to the customer, rather than in the depths of the central office. These systems support the provisioning of broadband voice and data applications, and they will help to realize the true promise of convergence.

A significant amount of work is underway to develop the vision of a truly capable optical metro solution. Manufacturers have realized that a metro aggregation device must have the following characteristics:

- The process of performing service aggregation must be collapsed into a single equipment layer. This not only saves on space, but also simplifies the management and provisioning processes.

- Optical transport with DWDM must be scalable but must respect the economics of the metro market while doing so. Service providers should be able to turn up new wavelengths only when necessary, thus preserving the efficiencies of the ring architecture. Chromatis, for example, a Lucent Technologies company, is designed in such a way

Figure 7-12
Complexity increases
toward customer.

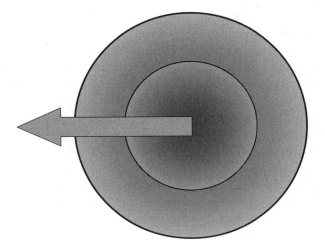

that service providers can turn up DWDM channels between ring nodes on an as-needed basis. In other words, DWDM can be added to individual nodes only without having to add it to the entire ring.

■ The system must enable end-to-end provisioning across the metropolitan area and mask the complexity of the underlying optical systems.

One key question that remains to be answered is this: Will fiber ever reach as far as the home or small office? Absolutely. However, certain barriers must first be overcome such as the cost of optoelectronics, the dearth of optical interfaces on consumer equipment, and network availability limitations. In the meantime, alternative solutions have emerged with varying degrees of success, such as *Hybrid Fiber/Coax* (HFC) architectures that take advantage of in-place, fully functional wiring schemes. HFC is considered to be a relatively low-cost solution for extending the reach of fiber. In the last few years, however, a number of fiber-to-the-home projects have been initiated worldwide, providing optical local loop connectivity to more than 300 million potential lines. SBC Communications continues to develop its broadband access plans, with plans to spend more than $6 billion on its broadband infrastructure, including Project Pronto, which includes 12,000 miles of metropolitan fiber to support the deployment of high-speed DSL service.

Metropolitan access is characterized by the deployment of ring architectures, used for the aggregation and transport of lower speed traffic. For example, a carrier might deploy a 10-Gbps metropolitan ring throughout a large city, as shown in Figure 7-13, which would then interconnect to lower speed, 2.5-Gbps access facilities—either point-to-point circuits or rings. SONET/SDH, as well as DWDM, are key technologies in this environment; ILECs/Incumbent PTTs and CLECs/City Carriers are involved in this segment of the marketplace, where they serve as peering points in the network, providing high-speed interfaces to multiple protocols, technologies, and companies.

Some service providers have reinvented themselves as broadband access carriers, offering a wide array of high-bandwidth access options including DSL, cable modems, and wireless solutions. These broadband access providers face a different set of issues than traditional carriers. First, they do not have a great deal of experience managing high-bandwidth access services and have never seen the tremendous growth that currently characterizes the market. Second, they tend to be quicker and more nimble than their more traditional competitors, making technology decisions that are in the best interest of their customers based on economies of scale and the potential to generate added revenue in innovative ways. These companies must deploy the most current technologies and must ensure scalability if they are to meet the growing demands of their customer base.

Figure 7-13
Backbone feeder
application.

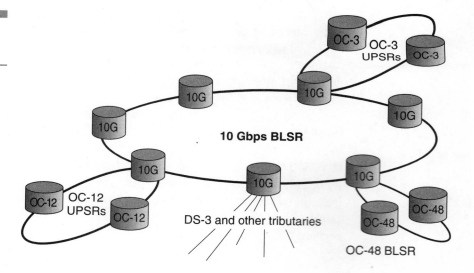

Many believe that these companies may eventually "own" the customer as broadband access catches on. They are deploying high-speed architectures designed to support the requirements of telecommuters, remote office installations, wireless business access, Gigabit Ethernet interoffice communications, and regional ISPs.

For example, a carrier might deploy an OC-48/STM-16 metro ring that provides transport for traffic that originates on DSL-equipped local loops, interconnecting remote workers with a corporate network elsewhere in the metro area. Another carrier might deploy an all-optical metro infrastructure to support the huge traffic increases between base stations that result from the deployment of third-generation wireless services. As data access over cable becomes more common, a carrier might build a high-speed access network designed to aggregate and transport traffic between cable customers and an ATM backbone.

As wireless local loop technologies such as the *Local Multipoint Distribution Service* (LMDS), shown in Figure 7-14, find their way into the business access domain, broadband metro carriers will roll out high-speed rings to satisfy the demands of these and other similar services. Similarly, a regional ISP, with a need to connect to multiple *Network Access Points* (NAPs) and database locations, may be served by a broadband metro carrier's 10-Gbps metropolitan ring, which provides transport for traffic that originates on low-speed services such as OC-48/STM-16, OC-3/STM-1, and traditional TDM services. Finally, a metropolitan ring can be used to interconnect corporate *local area networks* (LANs). Several vendors have deployed multiprotocol multiplexers that enable Gigabit Ethernet to be transported across a 10-Gbps ring—clearly an application with promise, given the widespread deployment of Ethernet technology. This is illustrated in Figure 7-15.

Metropolitan transport is precisely what the name implies: that segment of the transport market that delivers the high-speed rings used to aggregate and move lower speed traffic between locations or onward to a wide area transport environment. Finally, metropolitan enterprise is the realm of innovative access techniques designed to provide high-bandwidth solutions for businesses.

Figure 7-14
LMDS as an access technology.

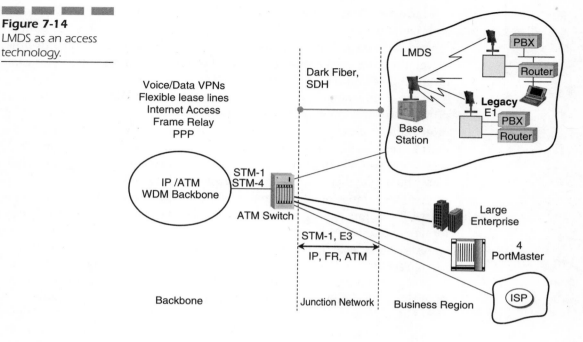

Figure 7-15
Metro Ethernet transport.

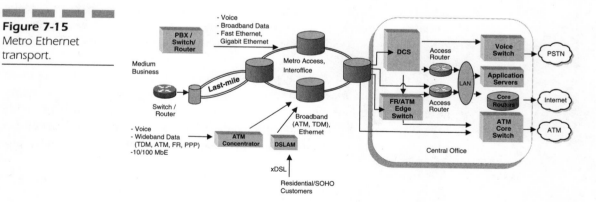

Packet and Other Data Services

Today's service provider networks designed to transport IP traffic or other data services must provide QoS that approaches that of legacy voice networks. Long ago (3 years!) we realized that the "killer application" for voice networks was data—that is, the addition of data-based services converted voice, traditionally a flat revenue service, into a service that created significant additional revenue though Caller ID and other data add-ons. Today, we have come to realize that the killer application for data networks is voice, the result of which is a clear recognition of the importance of survivable, dependable network infrastructures.

In response to this realization, service providers are merging their existing voice and data networks into a single multiservice infrastructure that matches the "five nines" of reliability and availability of the voice network, rather than the far less stringent standards of data networks. Five nines of reliability means that the network or component is 99.999 percent available.

Equally important is the concept of protocol agnosticism. Common questions are, IP or ATM? ATM or frame relay? MPLS or some other QoS-aware protocol? Each of these has value when deployed for specific types of services. For example, ATM provides the QoS control that is required for voice and private line services, whereas IP does not yet do so. Rather than coerce service providers to choose one protocol over another and compromise delivered service quality or to maintain multiple networks for different services, evolving data-oriented technologies enable the creation of a single protocol-independent infrastructure that can switch any protocol anywhere, anytime, with appropriate QoS for each service. The bandwidth capabilities of the optical network add greatly to this capability.

Another ongoing evolutionary requirement of modern networks is a recognition of the fact that the core and the edge are becoming functionally independent in many ways yet are functionally merging in others. The current layer-2 (ATM) and layer-3 (IP) multiservice networks are divided between the core and the edge. The edge, where IP first appears and QoS is introduced, provides service delivery, while the core, with its high-bandwidth fabric and QoS-aware switches, aggregates those services for transport. In reality, current service provider networks have three functional regions as shown in Figure 7-16: service creation and delivery at the edge, typically performed by smaller multiservice switches and routers; the aggregation of services within the multiservice core fabric, typically performed by larger ATM switches or core IP routers that aggregate traffic from the edge; and finally transport, the process of moving through the optical network while preserving QoS.

Tomorrow's networks will evolve from three layers into two. Many network designers believe that one of the first evolutionary steps will be a merger between the edge and the multiservice core, characterized by devices that deliver end-user services and also provide low-speed transmission aggregation for handoff to the optical core. The benefits of such a model are numerous, including a reduction in network complexity, a simplification of management requirements, a large increase in available bandwidth, and on-demand provisioning of QoS-based customer services.

The requirements for this enhancement to the existing network are not overly significant. The aggregation currently performed by large access nodes with high bit-rate DWDM facilities that interface directly to the optical core will migrate to the smaller hybrid devices that straddle the line between the core and the edge. Additionally, traffic-engineering functions must be modified to accommodate the provisioning of a selective partial mesh fabric for backbone connectivity. This evolution is most likely a 1- to 2-year process for the more aggressive second-tier carriers, longer for the first-tier companies.

Figure 7-16
The network's three functional regions.

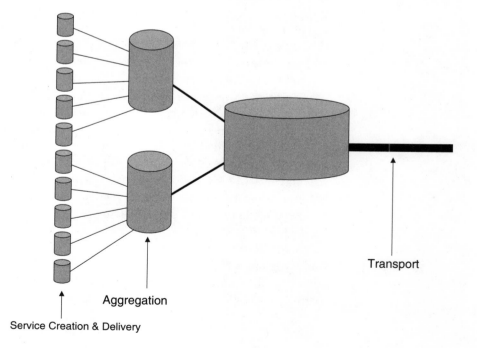

Transport

Aggregation

Service Creation & Delivery

And the benefits of this architectural modification? There are many, and they include the following:

- The capability to provide support for multiple services including ATM, frame relay, and MPLS-enhanced IP

- True, guaranteed, end-to-end, user-definable QoS, Per VC queuing, and QoS-aware routing

- Management control at the data link layer (layer 2) to improve the traditional *operations, administration, maintenance, and provisioning* (OAM&P) functions that are associated with the management of network resources

- Seamless network availability and flexibility through "lossless" switchover, automatic protection switching, and traffic management functions

- The capability to transport diverse services over a single, shared network infrastructure

- The capability to provide end-to-end, measurable QoS

- Support for an efficient mix of traffic types

- A flexible migration path with interfaces to existing DCS and SONET/SDH equipment

- The growth potential to deliver as-yet-unknown services

So how would these capabilities be realized in the network? The answer lies in the overall architecture of the network itself, specifically the optical core surrounded by a ring of layer 2 and 3 services.

Let's first consider the layer 2 and 3 service environment as having two major segments. The first is the multiservices ATM segment. The services that can be delivered from this segment include frame relay, ATM, MPLS+ (standards-based MPLS), private line, voice, and DSL.

Now consider the MPLS/IP segment. Services that can be offered from this segment include MPLS/IP and frame relay. What holds the two segments together is MPLS. In most cases, this is expected to be MPLS running over packets over SONET/SDH at OC48c/STM16 or OC192c/STM64.

As we observed earlier in the book, two key goals are associated with the deployment of telecommunications technologies: cost reduction and the creation of additional revenue.

Cost reduction can be accomplished readily with the deployment of new technology solutions, particularly in legacy data installations. Manufacturers have stepped up to this challenge by designing and manufacturing devices with characteristics that address the customer concerns that are frequently cited. These characteristics include the highest port density per installed chassis and multiservice support of the lowest lifecycle cost through the capability to add cards rather than change entire network elements as customer service requirements evolve. Additionally, these products take advantage of the strengths of high-speed layer 2 switching, offer a range of granular QoS levels, and provide the capability to deliver such IP-based applications as *virtual private networks* (VPNs), virtual routing capabilities, voice, fax, and video-over-IP, and IP Multicast.

Given that investments in legacy switching infrastructures continue, there is no question that they will remain a significant part of the evolving IP-over-ATM network. The evolution that most see is a slow but steady evolution from a circuit-based architecture to one based on packet switching, with ATM providing the capability to ensure QoS and toll-quality services. In effect, ATM will replace traditional "class four" tandem switches.

The other area of financial impact is revenue generation. New applications demand the shortest possible predictable end-to-end latency (40 microseconds), multiple priorities per installed port, and prioritized weighted fair queuing and scheduling to permit the delivery of predictable quality *voice over IP* (VoIP), video, and differentiated data services over an all-IP backbone. In this environment, offered applications include CBR over IP; voice, video, and virtual leased line services; guaranteed service-level agreements; enhanced services such as multicasting with multiple classes for real-time and non-real-time transport; and finally the capability to offer high degrees of network reliability and availability through the use of in-service software upgrades, circuit and device redundancy, and so on. All of these lend a degree of capability to modern data applications in the SONET and SDH realm.

ATM's Role

Of course, ATM's role in this evolving data network architecture is to enable convergence through the delivery of not only VoIP with guaranteed service quality levels, but also voice over frame relay, voice transport over ATM, and native voice traffic from the PSTN over an ATM infrastructure through the use of modern switching technologies that have recently emerged. Advancements are still required to effect the integration of signaling between the voice and data worlds, but this is an ongoing effort and a significant step toward the delivery of legitimately converged services.

What About Frame Relay?

Many service providers today offer frame relay to their enterprise customers for VPN support and leased-line alternatives. This has proven to be a popular alternative throughout much of the U.S. and is emerging as a viable solution in the rest of the world as well. There is, however, a drawback.

Today, because of their burgeoning data requirements, companies are beginning to exceed the capacity of traditional frame relay and are looking for more. Some advances have been made in frame relay environments including the capability to offer QoS and to create multilink, bonded frame circuits as a way to offer better bandwidth flexibility. Although these techniques are largely proprietary, they *can* be used to scale frame relay services, although both require a core network that provides higher throughput and therefore sustainable QoS. Once again, ATM rises to the challenge. With the publication and ratification of FRF.5 and FRF.8, the Frame Relay Forum put into place acceptable standards that govern the interworking of ATM and frame relay.

The end result of this effort is that manufacturers now have the ability to map frame relay QoS parameters into ATM cells in the core network, as shown in Figure 7-17. This enables the service provider to offer true end-to-end QoS and to be competitive with service-level agreements. This makes possible an enhanced revenue stream for service providers and enables them to charge premium prices for QoS-aware frame relay services.

Figure 7-17
Frame Relay
Bearer Service.

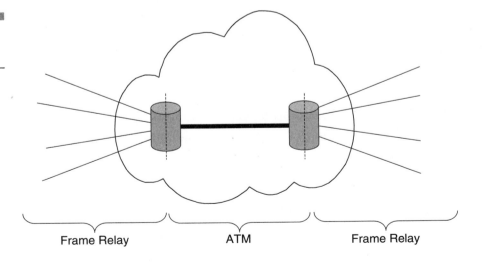

Wavelength-Based Services

Throughout all of these optical applications, a continuum of capabilities repeatedly appears. These networks have capabilities that fall into three identifiable categories: the characteristics of the network itself, the services that it provides, and the implications of those services. They have significant bearing on the deployment of wavelength-based services.

Network Characteristics   Networks that serve the needs of diverse customers with varying service requirements must exhibit the following characteristics. They must be scalable to meet changing bandwidth demands and anticipate unexpected growth. They must be reliable, offering five to seven nines of reliability at all times to emulate the service level provided by carrier class voice networks. They must be flexible, offering the capability to evolve rapidly in the face of changing customer requirements and traffic patterns and finally, they must be intelligent so that they can offer value-added services over a network infrastructure that is historically flat in terms of revenue generation.

Network Services   The network services straddle the gap between the characteristics that define the network and the services' implications. They include scalable, large-volume bandwidth, service assurance and guaranteed levels of customer service, capital and expense savings derived from increased operational efficiencies, and faster provisioning in response to customer requests for the same. These are the characteristics that network service providers can convert into a competitive advantage as well as competitive products and services.

Service Implications   Regardless of the measurable services deployed by a carrier, all have the following characteristics if they are successful:

- Fast and accurate response to the customer base and a subsequent fast time to market
- Better profit margins as the result of improved cost savings and more efficient network operations
- Improved revenues for a variety of related reasons
- The lowest cost per delivered bit of bandwidth

All three of these—network characteristics, network services, and service implications—play key roles in the quest for marketshare.

Modern Trends in Optical Networking

The numbers don't lie: bandwidth consumption is at an-time high and is going up, as Figure 7-18 illustrates. According to industry sources, consumption rates are off the charts. Every minute around the world, 5 million e-mail messages are transmitted. Every hour, 35 million voice mail messages are logged. Every day, 50,000 new wireless subscribers join the ranks of people trying to drive, fix a sandwich, read the stock picks, and talk on the phone at the same time. Every week, 630,000 new phone lines are installed, and every 100 days the total transmitted traffic on the Internet *doubles*. It gets better: the physical size of the Internet doubles every 10 months, and the amount of information that can be reached by typing "www-dot-something-or-other" doubles every 57 days. Furthermore, a collection of other factors doesn't help the situation at all. Every year, storage capacity doubles, while the price per unit sold halves, optical transport capacity doubles every 9 months or so, and Moore's Law, that wonderfully prophetic observation attributed to Intel's Gordon Moore, proves itself to be true every 18 months when the price halves and the speed doubles on computer CPU chipsets.

Figure 7-18
Exponential growth.

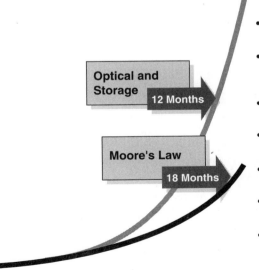

- Every minute, 5 million e-mail messages are sent
- Every hour, 35 million voice mail messages are left
- Every day, 50,000 people sign up for wireless adding to the 200 million worldwide
- Every day, 37 million log on to the Internet
- Every week, 630,000 new phone lines are installed
- Every 100 days, Internet traffic doubles
- Physical size of Internet doubles every ten months
- Volume of accessible data doubles every 57 days

All of these factors contribute to the massive challenge facing service providers today. There is an old adage about "fixing the train while it is rolling down the track" as a description of how to run a telephone company today. Pretty accurate, wouldn't you say?

The challenges before the service providers today fall into three categories: cost reduction, revenue generation, and meeting the demands of customers. Cost reduction can be achieved through a number of activities such as

- Eliminating the "network of networks" that most of them operate by moving to a single network (or at least a limited number of networks)
- Minimizing the overlay network buildouts that are so often required and therefore maximizing the capability to redirect those resources to more immediately profitable aspects of the business
- Simplification of the OAM&P functions in the network
- Preservation of the existing investment in the network through an "If it ain't broke, don't fix it" mentality (as long as it really isn't broken!)

Revenue generation, on the other hand, is a far more elusive beast. Clearly, competitive pricing is a critical factor and is directly affected by the items listed earlier. Others include

- The *absolute* ability to deliver guaranteed, always-there quality of service associated with purchased products
- The ability to act upon customer requests for service quickly and accurately and provision them in as short an installation interval as possible
- A collection of service offerings that are highly differentiated and "tailorable" to meet individual customer demands

Finally, companies that can meet the demand for bandwidth, particularly given the chaotic nature of the business today, will be among the most successful. Their requirements include

- The ability to support multiprotocol transport with complete agnosticism
- The ability to accommodate the unexpected service churn that is so much a part of doing business in the telecommunications marketplace today

- Support for unpredictable, constantly changing traffic patterns with widely varying requirements for QoS
- The ability to handle both anticipated and unexpected demands for additional capacity

Ultimately, wavelength-based services lend themselves perfectly to service providers faced with the challenges described. They help assuage the chaos brought about by unpredictable customer demands for bandwidth, provide virtually unlimited network scalability, enable dramatically reduced costs of operations, and offer the protocol transparency and broad flexibility required in modern networks. In the end, wavelength-based services offer true bandwidth on demand, protocol-independence, unlimited bandwidth provisioning on a per-wavelength basis, the capability to provision services quickly while offering granular control of QoS, and, perhaps most important of all, customer control over the process. Together these characteristics make for a powerful combination that solves concerns for service provider and customer alike.

Traditional Versus Wavelength-Based Services

When we consider the differences between the so-called legacy network installations and the modern wavelength-based services, the differences are striking. Consider the network diagram in Figure 7-19. In a traditional network, services are carried over a complex, difficult-to-manage and expensive overlay fabric that is highly bit rate-and protocol-dependent. Because of the patchwork nature of the network architecture, it requires multiple management systems, enjoys a time-consuming, error-prone provisioning process, and often has resources sitting idle because of underutilization.

Wavelength-based network provisioning systems, on the other hand, require fewer network elements, make much more efficient use of expensive network resources, and have the capability to deliver bandwidth on demand. Because of the distributed intelligence model that is so common in modern optical networks, wavelength-based network provisioning systems are also capable of provisioning client service requests far faster and with greater accuracy than their legacy counterparts. Furthermore, solutions

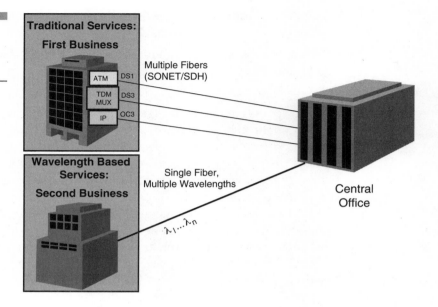

Figure 7-19
Traditional vs.
wavelength-based
services.

can be crafted that address such concerns as survivability, performance monitoring, and fault isolation.

All network topologies can be used in optical networks that deliver lambda-based services. Many carriers deploy dedicated facilities initially, simply because they are easy to manage, relatively low-cost, and designed to address specific customer concerns. Some build rings, although they are very expensive, the price that must be paid for their high level of reliability. Mesh installations are typically more cost-effective and are designed to make more efficient use of scarce and expensive network resources. Quite often it is the application that drives the desired network architecture; we will discuss applications for wavelength-based services in the following section.

Wavelength-Based Applications

Wavelength-based applications fall into four main categories today, although there will certainly be others over time as creativity and the quest for dollars continue unfettered. These four categories are bandwidth trading, optical VPNs, traffic aggregation and transport, and multi-domain WAN transport.

Bandwidth Trading

Bandwidth, along with Chicago pork bellies, Louisiana sweet crude, Costa Rican coffee, Kansas wheat, and Pacific Northwest lumber, is fast becoming the next big commodity market product, particularly when sold on a per-minute basis.

In bandwidth trading, companies like Enron, Williams, and Arbinet purchase large volumes of available bandwidth from providers with excess capacity, thus taking on the considerable risk of owning a commodity with no initial customer to buy it. Ideally, however, it all works out in the end.

How It Works

When a customer wants to buy a contract for bandwidth from a bandwidth trader, as shown in Figure 7-20, the customer contacts a trader/broker and explains what his or her capacity requirement is. The trader negotiates a contract between the customer and the provider of the bandwidth. The bandwidth trading application, hosted by the trader's company, requests a new service from the software program (sometimes called the service manager) that controls bandwidth provisioning. The service manager, in turn, contacts the so-called "pooling points" from which the bandwidth will be provisioned and monitors the delivered level of QoS at each of them.

Figure 7-20
Bandwidth trading.

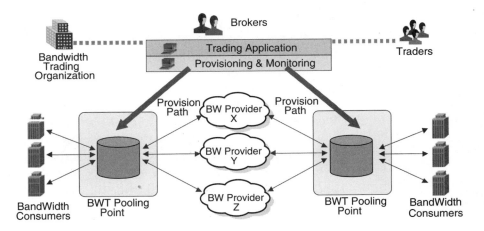

Companies like Enron have suggested that bandwidth, along with power, is the next product that will be widely sold in a forward commodity market. As one might suspect, however, this concept does not sit well with the legacy carriers, since the commoditization of bandwidth represents a direction in which they don't want to go. In spite of that, many service providers are customers of bandwidth traders, buying bandwidth from the cheapest source as a way to control their own network costs.

Enron trades as the principal player in the bandwidth commodity game, which means that the company pays for the products it is buying and selling, often before it has a customer. As the principal player in the transaction, Enron can play any number of roles: they can hedge, serve as the intermediary, or be a speculator. As a hedge, they can offset transactions as a hedge against less desirable positions. As an intermediary, they can position themselves between a facilities-based carrier such as an ILEC and a non-facilities-based carrier such as a CLEC, and buy bandwidth from one to sell to the other. As a speculator, Enron can either sell available bandwidth or buy excess capacity from another company, hoping that the price changes favorably to allow them to play in the arbitrage game and make a large profit.

To raise awareness of the potential value of bandwidth trading, and to increase interest in it among the carriers, both Enron and Williams petitioned the Competitive Telecommunications Association (CompTel) to create an ad hoc working group to examine the possibility and feasibility of creating a bandwidth-trading organization that would push the industry for a standardized trading agreement that could be used universally. In 1999, CompTel created the Bandwidth Trading Organization (**www.bandwidthtradingorganization.com**), and although no formal agreement has yet been forged, a great deal of work has been done, including the publication of a mission statement and the circulation of a straw agreement among all players. A formal trading agreement is expected soon and will only address the provisioning of SONET/SDH circuits. In the future, as provisionable IP circuits gain capacity and standards for provisioning wavelengths and wavelength interfaces become more common, commodity trading of those resources will become more common.

Challenges still face the market for commoditized bandwidth, although they are not show-stoppers by any means. Until pooling points become more common and a standardized provisioning contract is created, freely operated markets for bandwidth will still stumble a bit.

The energy consortia are closer than the telecom companies to establishing a standard contract, but this will turn around in the relatively near future. The greatest stumbling block in the telecom arena is the legacy mindset that is so infused with risk-averse behavior. Carriers want the most for their investment dollar and are not accustomed to operating in a marketplace that is fraught with some degree of risk. Their greatest revenues come from value-added services, and as a result, they are loathe to endanger that revenue stream. They are justifiably concerned that product standardization would lower prices, damage their current product differentiation, and cause the creation of a commodity market.

Of course, certain carriers support the success of bandwidth trading, among them the more enlightened companies that understand the value that the process can bring. They also recognize that it is inevitable: bandwidth has all the characteristics of an emerging commodity product, and it is therefore only a matter of time before it achieves full-blown soybean status. Transactions for bandwidth purchases that used to take months to complete will be completed in minutes or seconds, causing the wholesale market efficiencies commonly seen in other industries to appear in telecommunications.

Many energy companies like Enron and Williams operate pipelines for oil and natural gas and therefore own significant rights-of-way, along which they can install fiber optic infrastructures. By using these rights-of-way, these companies avoid a significant cost.

Enron has approximately 18,000 miles of fiber installed in its network and was the first company to treat bandwidth as a commodity. According to company sources, Enron's telecommunications organization estimates the current market for commodity bandwidth in North America alone to be in excess of $300 billion, $1 trillion worldwide. Those are not numbers to ignore.

Enron has not waited for the CompTel Bandwidth Trading Organization to come up with its standards. The company has developed standardized terms and conditions for the movement of commodity bandwidth. It has also designed and proposed an operations model for universal pooling points that would serve as interconnection facilities in certain markets through which connections between buyers and sellers could be created. An independent third party responsible for scheduling connections and ensuring physical and logical transaction security would most likely manage them.

Optical VPNs

Because the virtual corporation comprises distinct and often geographically dispersed locations, and because remote workers and home offices have become so common, networks have evolved to support interconnections of multiple offices at a reasonable cost and remote access to corporate databases. The competitive nature of today's marketplace requires that employees have immediate and efficient access to the most current information available, and that they be able to work remotely as efficiently as if they were directly connected to the corporate network. Another advantage deals with extending the cloud beyond its traditional margins. By creating a virtual corporate network, customers can be "brought in" electronically to ensure the timely exchange of information and enhance customer relationship management. One solution that is commonly used is the Virtual Private Network (VPN), which is a cost-effective alternative to a dedicated facility. Instead of paying the mileage charges associated with private line, a VPN replaces the dedicated circuit with the public Internet, the use of which has no distance component associated with it.

When a remote user accesses a corporate network using a VPN, they do not dial directly into the corporate network. Instead, they create an Internet connection by dialing into a local ISP and then use secure protocols to create a "tunnel" through the fabric of the Internet that safely transports the information between the remote worker and the corporate network. The Internet simply becomes an extension of the corporate network, providing access for remote workers. To protect the user and the company, secure protocols enable information to be passed safely between a remote employee and corporate computer systems. The difference between these two techniques is shown in Figure 7-21.

Design considerations must be taken into account when implementing a VPN. If the majority of the users are geographically collocated, a dialup solution may be perfectly acceptable since they will incur no distance-related charges as a result of their connection sessions. If they are more geographically dispersed, however, an Internet or alternative IP solution may be a better answer.

VPNs represent a secure and cost-effective alternative to a dedicated network, making it possible to interconnect geographically dispersed corporate LANs and WANs using a low-cost public network infrastructure. Currently, private line and frame relay represent the bulk of all WAN installations. Unfortunately, they are relatively inflexible, a problem in

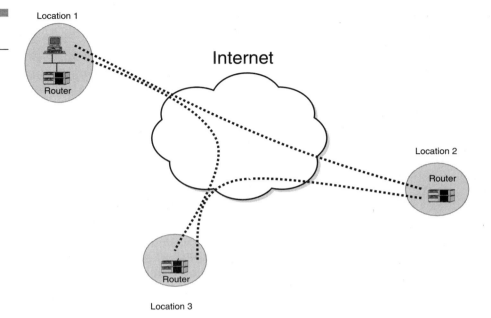

Figure 7-21
Traditional VPN.

today's dynamic network environment. And because they are usually sold under a 3-to-5 year contract, they have become less attractive for customers who do not want to be locked into a network solution that cannot evolve with changing applications and geographic footprint requirements.

The cost advantages of VPNs are significant. In addition to the savings that result from the elimination of distance charges and the costs of dial access, there are also savings from reductions in capital equipment and support. Because VPNs use a single WAN interface for multiple functions, the data that would normally have passed through several devices now requires one. Support costs are reduced because of the capability to consolidate all support functions within a single help desk organization.

Of course, other challenges are associated with VPNs, not the least of which is performance. The Internet is not known for providing highly dependable QoS, and given that it constitutes the heart of the VPN, a number of questions arise. An ideal IP-based VPN would

- Be ubiquitously available, secure, and reliable
- Offer a variety of network management and billing options

- Provide measurable service level agreements
- Allow the user to differentiate QoS on a per-flow and/or per-application basis

The Internet, or even a pure IP network, does not inherently have the capability to do all of these things. IP on top of an ATM backbone, however, does have the capability, and this model seems to be the emerging choice for carriers looking to deploy QoS-capable public networks. By associating IP addresses to ATM virtual circuits, QoS can be assured.

Along Comes the Optical VPN

The next stage in the development of the VPN is the addition of optical technology to the transport solution. Optical VPNs add the advantage of extremely high bandwidth as well as the following:

- The capability to provision multiple security levels using any variety of techniques
- Wavelength-based provisioning for complete control of bandwidth
- Customer control of the provisioning process, if the carrier desires
- GUI-based customer interfaces
- The capability to generate customized reports for clients

In an optical VPN, shown in Figure 7-22, the wide area optical network provides the high-bandwidth interconnection fabric for multiple company locations, all under the control of a network management system that tracks bandwidth consumption and customer requests for service, provisions as appropriate, and manages billing and accounting functions automatically. The same secure protocols used in the traditional VPN can be used in the optical VPN.

It is important to recognize that the private line network will never disappear, but as Internet QoS and encryption protocols improve, and optical fiber becomes more available for transport, the need for a dedicated facility will become less and less obvious. Because of its capability to reduce costs, create customer controllable networks, support new business opportunities, and improve flexibility and speed to market, the optical VPN will be recognized as a powerful enabler of customer relationship management and therefore competitive positioning.

Figure 7-22
Optical VPN.

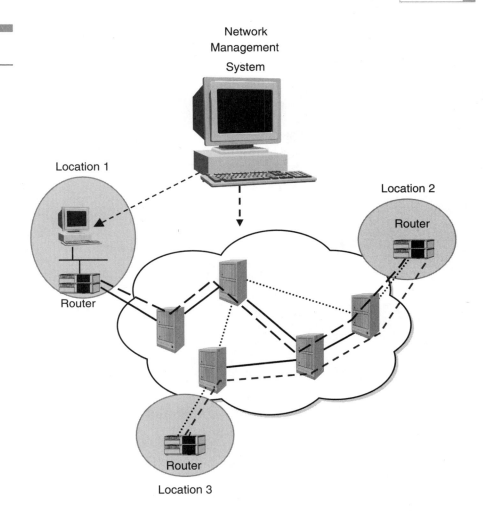

Traffic Aggregation and Transport

As corporations evolve such that they have multiple corporate locations scattered across a large geographic area, as shown in Figure 7-23, traffic patterns usually emerge between the various sites that are relatively predictable. If large volumes of traffic can be predicted between any two sites, then it makes sense to establish a wavelength dedicated for the transport of traffic between them. This technique, often called express traffic

Figure 7-23
Wide area, multi-
lambda transport.

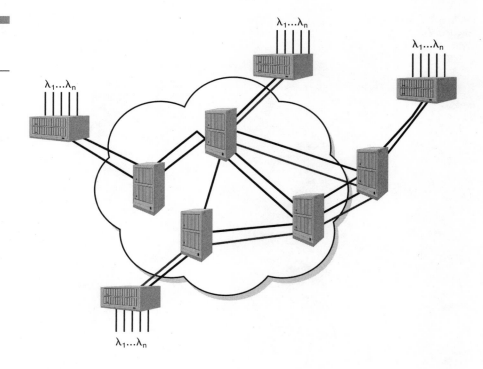

handling, is used to aggregate and transport traffic at high bandwidth levels between sites in a real or virtual corporate network. As Figure 7-23 shows, the network is designed with high-bandwidth junction points that provide cross-connect capability. Routing between sites is based on the nature and destination of aggregated traffic rather than on the management of individually provisioned wavelengths.

Multi-Domain Wide Area Network Transport

Perhaps a simpler name for this application would be "Submarine transport." This application is used when SONET and SDH traffic must be interconnected to form a single, logical network capable of moving the two traffic types seamlessly back and forth. Because of the gravity of the installation and the large volume of traffic that it will naturally carry, it must support a variety of restoration schemes to protect user traffic. It must also support rapid provisioning options.

Ethernet Transport

A fundamental disconnection exists between the core and the metro regions of the network that relates to the basic service offerings of each. In the metro domain, customers require the network to carry the following:

- DS-1, E-1, DS-3, and other legacy signals
- Frame relay and ATM traffic
- DSL and cable modem traffic
- Leased wavelength traffic
- ESCON and FICON signals
- Video
- Transparent LAN service traffic
- Ethernet, Fast Ethernet, and Gigabit Ethernet

The core network, on the other hand, is typically far less diverse in its design, carrying only packet over Lambda/wavelength, packet over SONET/SDH, and straight SONET and SDH traffic at 2.5 or 10 Gbps.

Matching the two environments is a formidable challenge for service providers who must meet the demands for growing bandwidth, evolving QoS-sensitive services, and flexibility. Ethernet, Fast Ethernet, and Gigabit Ethernet represent the fastest growing segment of metro-to-wide-area traffic and as such have become a primary area of focus for manufacturers of devices that straddle the metro/core interface.

The traditional central office that serves the metropolitan area, shown in Figure 7-24, is relatively straightforward in its design and mission. Traffic arrives on an optical facility, possibly over DWDM. The traffic is divided and passed into SONET or SDH multiplexers, where each traffic stream is then passed on to an ATM switch or digital cross connect system in the case of traditional voice services. Some of the traffic may be passed on to a router if it is IP-based.

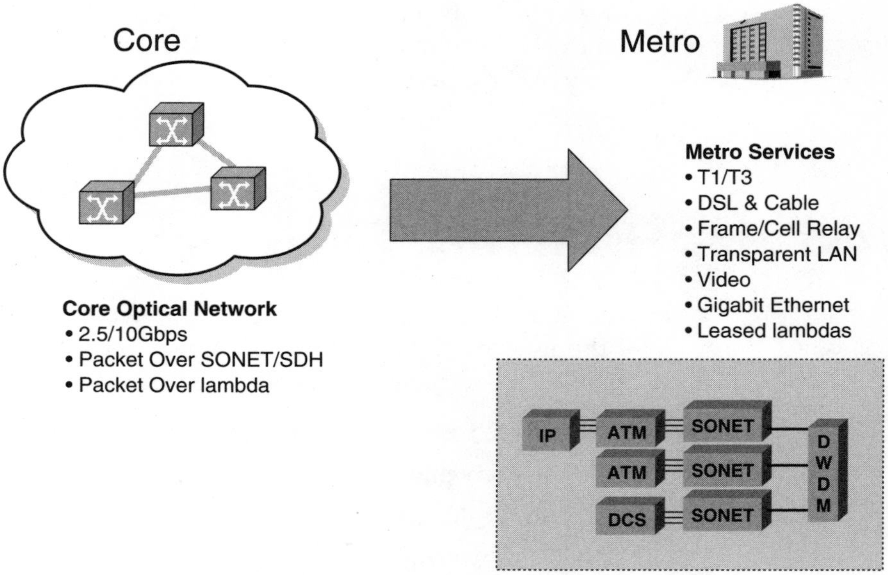
Core
Metro
Metro Services
• T1/T3
• DSL & Cable
• Frame/Cell Relay
• Transparent LAN
• Video
• Gigabit Ethernet
• Leased lambdas
Core Optical Network
• 2.5/10Gbps
• Packet Over SONET/SDH
• Packet Over lambda
IP
ATM
ATM
DCS
SONET
SONET
SONET
DWDM

Figure 7-24 The traditional metro office.

In the evolving metro central office, three techniques exist for handling traffic. The first is with an all-optical design, in which case individual wavelengths are assigned to specific protocols, as shown in Figure 7-25. The second technique is the so-called next-generation SONET/SDH environment, illustrated in Figure 7-26. In this environment, the cross-connect function and the add-drop function are integrated into a single device. Lower-speed services arrive at the DCS and are then mapped into outbound SONET or SDH channels for wide-area transport. The final technique, shown in Figure 7-27, is the multi-service model. Here scalable DWDM optics are interfaced to the ADM function, which in turn is connected to multiple service elements that provide ATM, frame relay, IP, Ethernet, and other desired services.

The result of this overall design variety is that networks can be deployed that effectively provide transport for Ethernet and other traffic types between the core and metro environments.

Figure 7-25
All-optical CO.

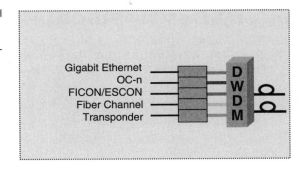

Figure 7-26
Next-generation
SONET/SDH CO.

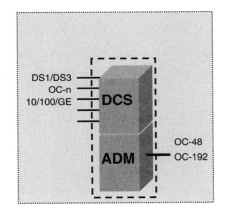

Figure 7-27
Multi-service CO.

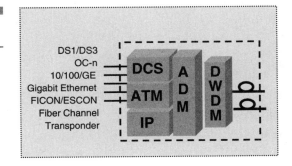

Corollary Developments

One of the most innovative technologies to come along in recent years is *passive optical networking* (PON), shown in Figure 7-28. PON comprises nothing more complex than an optical splitter that sits between the customer "loops" and the transport fabric and creates passive optical channels that are individually fed into the transport network. PON can be used very effectively to transport legacy voice, ATM traffic, and all flavors of Ethernet traffic. When combined with ATM on the backbone, it provides a technique for the fiber-based transport of voice and data services that originate at small and medium-size businesses. The advantages of *ATM PON* (APON) are numerous:

■ Fully compatible with the service provider's ATM network backbone

■ The capability to provision bandwidth on-demand

■ A highly competitive solution that is far more cost-effective than SONET/SDH

Figure 7-28
Passive optical networking.

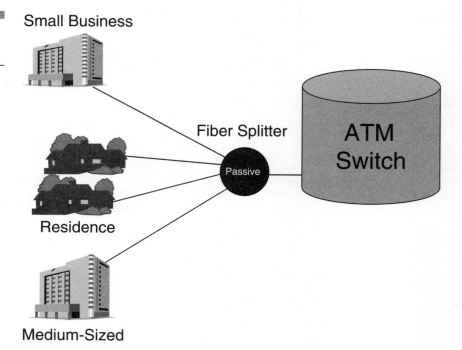

- A solution that results in significantly lower hardware costs because of the advantages of convergence
- Reduced cost per subscriber because of shared access to a network host
- Based on a well-accepted set of international standards
- Supports ATM, Ethernet, and circuit-based services

Because Ethernet has become the primary protocol used in corporate environments, the demand to transport it across metro and WANs has grown significantly.

Consider, for example, the services offered by Yipes. Their network architecture converges data, voice, and video over an optical networking infrastructure to satisfy the demands of e-commerce, Internet-based enterprises, and multimedia delivery. These services can only be addressed with optical IP networks. The challenge for Yipes has been to meet customer requirements for guaranteed minimum bandwidth to ensure the delivery of the QoS required for multimedia and latency-sensitive applications such as Voice-over-IP.

Yipes uses Ethernet to transport IP over optical fiber. This combination offers advantages over legacy WAN architectures in terms of reduced cost, simplified network design, scalable bandwidth, and time to market without reducing reliability or QoS. Bandwidth is delivered through a standard RJ-45 port that enables the delivery of Internet connectivity, point-to-point services, multipoint-to-multipoint services, and VoIP at bandwidth levels ranging from 1 Mbps to 1 Gbps. The network, which is entirely non-blocking, guarantees extremely low latency and is thus acceptable for delay-sensitive traffic. Architecturally, redundant fiber rings ensure reliability and availability.

Another company in the Ethernet transport game is Telseon. Telseon delivers secure and scalable bandwidth across their own optical network at speeds up to 1 Gbps. They rely on standard Ethernet connections, which eliminates customer dependence on expensive high-speed data interfaces and the management overhead required to deploy and maintain a legacy SONET or SDH network. The company's customers include

- IP backbone providers
- Storage area network (SAN) providers
- Content providers and distributors
- Managed service providers

- Application service providers (ASPs)
- Web enterprises

Telseon recently made news in the trade journals when it declared itself to be the first bandwidth provider to place provisioning functions in the hands of customers. This allowed them to self-provision virtual connections and gave them the capability to modify bandwidth as required in real-time. Telseon also offers business-to-business connectivity, which enables any company on Telseon's metro network to create a connection and offer it to any other company on the company's network. Once network access has been established, Telseon enables customers to add and change connections and increase bandwidth on demand as they require, all using Telseon's Web-based IP provisioning system.

The company offers three levels of service:

- *Point-to-point.* Connects two sites, either within your own network or from your network to a service provider or business partner network.
- *Point-to-multipoint.* Service providers connect to multiple customer sites or for multicast distribution.
- *Multipoint-to-multipoint.* A single organization connects to campuses, multiple carrier POPs, or trading communities.

The Multiservice Optical Core

The multiservice optical core network looks like one of those Russian mamuschka dolls—a series of nested service layers that together create a multiservice network. This network is designed to transport a wide variety of service types, including

- ATM
- MPLS-enhanced IP
- DSL
- Frame relay
- Packetized voice
- Cable modem traffic

- Private line
- Wireless
- VPN services

The network is most likely deployed over ATM as the switching infrastructure. This results in a number of advantages, including low latency/delay characteristics for transported traffic, the capability to prioritize, and enhancements to an otherwise all-IP infrastructure.

All of the services listed here benefit from the modern core design. The advantages are as follows:

- *ATM.*   Multiprotocol transport support, QoS preservation, and globally-accepted standards
- *MPLS-enhanced IP.*   Extremely efficient use of sometimes scarce addressing space, reuse of existing routing protocols, and scalability
- *DSL.*   High-bandwidth access to both intranet and Internet services over existing copper facilities and traffic aggregation over ATM access and backbone services
- *Frame relay.*   Reliance on standards-based *frame relay bearer service* (FRBS) and the ability to offer end-to-end seamless QoS as a competitive advantage
- *Packetized voice.*   Low-delay architecture associated with ATM that enables voice services to be given highest priority service and permits planning for circuit-to-packet migration strategies
- *Cable modem traffic.*   Capability to deliver granular QoS, capability to offer and provision high-bandwidth virtual circuit service, and makes high-speed Internet access possible
- *Private line.*   Multiprotocol transport and differentiable QoS
- *Wireless.*   Similar to packet voice with low delays, scalability, and high-quality predictable transports
- *VPN services.*   Efficient use of private and public IP addressing space, reuse of existing routing protocols, the capability to create "custom" address plans, and scalability

Optical Area Network

The *optical area network* (OAN) is the name given to the optical network infrastructure that provides gradable transport for a wide array of services including

- *Storage area networking* (SAN)
- Web content hosting
- Bandwidth trading (discussed earlier)
- Video services
- Data center connectivity

The reader should be able to see that these services are entirely content-driven and are therefore highly dependent on varying degrees of QoS. The deployment of these services is causing a massive reinvention of the network. The number of network users is climbing rapidly. Applications are being created that require more and more storage and transport bandwidth and that are often outsourced (which puts an additional burden on the network). Web pages are becoming far more graphics-intensive and therefore bandwidth-hungry, and the Web is becoming the delivery medium for a volley of unanticipated services including television, video on demand, MP3 music, soccer games, and more. The end result of this is that person-to-person communication is no longer the most common model for traffic. Instead, the model is rapidly becoming person-to-server, server-to-server, and data center-to-data center. The consequences of this evolving service infrastructure are significant, particularly for the data center environment (which in effect includes the SAN, Web hosting, bandwidth trading, and some video content environments).

Consider the characteristics of the typical modern data center today. They have evolved from centralized, hierarchical roots and are saddled with the vestiges of older technologies. They house large numbers of services, all interconnected with standard 10-Mbps Ethernet. They provide support and sustenance for LANs, SANs, and WANs. They support a variety of less-than-efficient legacy WAN interfaces that are left over from earlier times but that are deeply embedded and therefore cannot be removed. They are forced to run and support multiple redundant software architectures, and finally they must manage multiple database farms that do not communicate to one another.

The fact of the matter is that various realities are at work here that occasionally clash with one another. Within the data center, terabytes of content are stored in massive disk farms. On the WAN side, terabits of transport bandwidth are provisioned for the transport of data center content. Both of these represent significant costs, but the revenues from sold bandwidth and hosted data center content (hopefully) offset the costs. The OAN represents the solution that sits between the data center and the wide area, delivering scalable bandwidth that offers QoS, protocol agnosticism, and scalable bandwidth.

Conclusion

Throughout this final section we have discussed the applications that are emerging under the domain of SONET and SDH. These applications derive from a number of technology sources including native mode ATM, frame relay, private line data circuits, virtual private networks, DWDM-dependent wavelength-based services, and various flavors of Ethernet, IP, video, and wireless. The conclusion that is inevitably reached is multifaceted and is described here.

Long live packet switching. There is an undeniable evolution underway during which circuit switching's influence is waning, while that of packet switching is waxing. The reasons for this evolution include more efficient use of network resources, the capability to converge multiple services on a single network fabric, and competition.

Convergence is real. Get over it. It really is happening: IP is ascending to the throne of network layer protocols and *will* become the anointed protocol for routing purposes. All transported services are in various stages of reinventing themselves for transport across packet-based infrastructures.

QoS is the most important differentiator. It isn't bandwidth, availability, age, or fancy names. Customers buy services because the services do what they need to do—period. Companies that take the extra effort to understand customer requirements and customize service offerings to address the specific needs of each client's situation will ultimately win the game.

Content is the real driver. The realization that customer-to-server and data center-to-data center traffic is eclipsing the traffic generated by point-to-point connectivity highlights the fact that content rules. Whether it's interactive video, MP3 music, live sports events, BBC music, medical imaging, or some other form of content, it's the real driver behind the bandwidth explosion.

SONET and SDH are here to stay for the time being. They may have been around for a while, and they may not be the most flexible and accommodating protocols available, but they are deeply embedded in the network, offer a wide array of capabilities, are remarkably forgiving, and they work. Add to that list the fact that service providers know and understand them and you will see why revenues from these technologies will continue to grow for some time to come.

Optical networking is the next big hero. It's true: optical, with its virtually unlimited bandwidth, ubiquity, protocol transparency, and vendor support, is in the game for the long haul. There is nothing else on the horizon that stands a chance of unseating it in the near term.

Metro is the focal point for the foreseeable future. Currently, the area of greatest growth in the optical networking domain, the metro region, has long been somewhat ignored because SDH and SONET were simply not the best protocols to use there. Today, however, given the growth that is underway in the metro world and the bridging that must inevitably occur between the metro and core/long-haul domains, metro is becoming a market segment to be paid attention to.

Ethernet has escaped from the LAN and is roaming free. Once considered to be an almost hobby-like protocol, Ethernet has become the darling protocol for office interconnection and, if you believe the tales of Yipes and Telseon, an ideal IP transport across the wide area as well. Who woulda thunk it.

IP just isn't ready for primetime. It's a great protocol with tremendous promise and absolutely *will* become the protocol of choice for mixed service networks. Until it has the ability to deliver discernible, predictable, and controllable QoS, it will be required to work hand-in-glove with a robust switching fabric.

Rings are making room for meshes. Ring architectures are fine for the requirements of the metro market where they are widely deployed, but they are giving way to mesh network designs in the long-haul and core domain.

We have more than enough bandwidth—for now. Parkinson was right when he wrote, "Work expands to occupy the time allowed for it." A corollary to his now-famous law might be, "Applications will always expand to occupy the bandwidth allotted to them." Today optical channels on the bottom of the Pacific Ocean are lying idle, but that condition is purely momentary.

So what does all this mean? The technologies that made up the legacy networks of yesterday and the services that they provided have changed dramatically. Instead of the relatively inflexible and bandwidth-limited service provider networks being the model for transport today, they are giving way to converged networks that provide flexible connectivity, vast bandwidth reservoirs, highly survivable architectures, and protocol agnosticism. Some of those legacy technologies, however, specifically the high-bandwidth transport schemes of SONET and SDH with their enormously capable overhead bytes, have survived the technology purge and will continue to be significant players in the evolving broadband network.

GLOSSARY

A

Absorption   A form of optical attenuation in which optical energy is converted into an alternative form, often heat. Often caused by impurities in the fiber; hydroxyl absorption is the best-known form.

Acceptance Angle   The critical angle within which incident light is totally internally reflected inside the core of an optical fiber.

Add-Drop Multiplexer (ADM)   A device used in SONET and SDH systems that has the capability to add and remove signal components without having to demultiplex the entire transmitted transmission stream, a significant advantage over legacy multiplexing systems such as DS3.

Aerial Plant   Transmission equipment (including media, amplifiers, splice cases, and so on) that is suspended in the air between poles.

Amplifier   A device that increases the transmitted power of a signal. Amplifiers are typically spaced at carefully selected intervals along a transmission span.

Amplitude Modulation   A signal encoding technique in which the amplitude of the carrier is modified according to the behavior of the signal that it is transporting.

Analog   A signal that is continuously varying in time. Functionally, analog is the opposite of digital.

Angular misalignment   The reason for loss that occurs at the fiber ingress point. If the light source is improperly aligned with the fiber's core, some of the incident light will be lost, leading to reduced signal strength.

Armor   The rigid protective coating on some fiber cables that protects them from crushing and from chewing by rodents.

Application-Specific Integrated Circuit (ASIC)   A specially designed IC created for a specific application.

Asynchronous   Data that is transmitted between two devices that do not share a common clock source.

Asynchronous Transfer Mode (ATM)   A standard for switching and multiplexing that relies on the transport of fixed-size data entities called cells which are 53 octets in length. ATM has enjoyed a great deal of attention lately because its internal workings enable it to provide quality of service (QoS), a much-demanded option in modern data networks.

ATM Adaptation Layer (AAL)   In ATM, the layer responsible for matching the payload being transported to a requested quality of service level by assigning an AAL type that the network responds to.

Attenuation   The reduction in signal strength in optical fiber that results from absorption and scattering effects.

Avalanche Photodiode (APD)   An optical semiconductor receiver that has the capability to amplify weak, received optical signals by multiplying the number of received photons to intensify the strength of the received signal. APDs are used in transmission systems where receiver sensitivity is a critical issue.

Axis   The center line of an optical fiber.

B

Backscattering   The problem that occurs when light is scattered backward into the transmitter of an optical system. This impairment is analogous to echo, which occurs in copper-based systems.

Bandwidth   The range of frequencies within which a transmission system operates.

Baud   The *signaling rate* of a transmission system. This is one of the most misunderstood terms in all of telecommunications. Often used synonymously with bits-per-second, baud usually has a very different meaning. By using multibit-encoding techniques, a single signal can simultaneously represent multiple bits. Thus the bit rate can be many times the signaling rate.

Beam Splitter   An optical device used to direct a single signal in multiple directions through the use of a partially reflective mirror or some form of an optical filter.

Bell System Reference Frequency (BSRF)   In the early days of the Bell System, a single timing source in the Midwest provided a timing signal for all central office equipment in the country. This signal, delivered from a very expensive cesium clock source, was known as the BSRF. Today, GPS is used as the main reference clock source.

Bending Loss Loss that occurs when a fiber is bent far enough that its maximum allowable bend radius is exceeded. In this case, some of the light escapes from the waveguide resulting in signal degradation.

Bend Radius The maximum degree to which a fiber can be bent before serious signal loss or fiber breakage occurs. Bend radius is one of the functional characteristics of most fiber products.

Bidirectional A system that is capable of transmitting simultaneously in both directions.

Bragg Grating A device that relies on the formation of interference patterns to filter specific wavelengths of light from a transmitted signal. In optical systems, Bragg Gratings are usually created by wrapping a grating of the correct size around a piece of fiber that has been made photosensitive. The fiber is then exposed to strong ultraviolet light that passes through the grating, forming areas of high and low refractive indices. Bragg Gratings (or filters, as they are often called) are used for selecting certain wavelengths of a transmitted signal, and are often used in optical switches, DWDM systems, and tunable lasers.

Broadband Historically, broadband meant "any signal that is faster than the ISDN Primary Rate (T1 or E1)." Today, it means "big pipe"—in other words, a very high transmission speed.

Buffer A coating that surrounds optical fiber in a cable and offers protection from water, abrasion, and so on.

Building Integrated Timing Supply (BITS) The central office device that receives the clock signal from GPS or another source and feeds it to the devices in the office it controls.

Butt Splice A technique in which two fibers are joined end-to-end by fusing them with heat or optical cement.

C

Cable An assembly made up of multiple optical or electrical conductors, as well as other inclusions such as strength members, waterproofing materials, armor, and so on.

Cable Assembly A complete optical cable that includes the fiber itself and terminators on each end to make it capable of attaching to a transmission or receive device.

Cable Plant   The entire collection of transmission equipment in a system, including the signal emitters, the transport media, the switching and multiplexing equipment, and the receive devices.

Cell   The standard protocol data unit in ATM networks. It comprises a 5-byte header and a 48-octet payload field.

Cell Loss Priority (CLP)   In ATM, a rudimentary single-bit field used to assign priority to transported payloads.

Cell Relay Service (CRS)   In ATM, the most primitive service offered by service providers, consisting of nothing more than raw bit transport with no assigned AAL types.

Center Wavelength   The central operating wavelength of a laser used for data transmission.

Central Office   The central switching facility where most voice and data circuits terminate.

Central Office Terminal (COT)   In loop carrier systems, it is the device located in the central office that provides multiplexing and de-multiplexing services. It is connected to the remote terminal.

Chirp   A problem that occurs in laser diodes when the center wavelength shifts momentarily during the transmission of a single pulse. Chirp is due to instability of the laser itself.

Chromatic Dispersion   Because the wavelength of transmitted light determines its propagation speed in an optical fiber, different wavelengths of light will travel at different speeds during transmission. As a result, the multiwavelength pulse will tend to spread out during transmission, causing difficulties for the receive device. Material dispersion, waveguide dispersion, and profile dispersion all contribute to the problem.

Circuit Emulation Service (CES)   In ATM, a service that emulates private line service by modifying (1) the number of cells transmitted per second and (2) the number of bytes of data contained in the payload of each cell.

Cladding   The fused silica coating that surrounds the core of an optical fiber. It typically has a different index of refraction than the core, causing light that escapes from the core into the cladding to be refracted back into the core.

Coating   The plastic substance that covers the cladding of an optical fiber. It is used to prevent damage to the fiber itself through abrasion.

Coherent A form of emitted light in which all the rays of the transmitted light align themselves on the same transmission axis, resulting in a narrow, tightly focused beam. Lasers emit coherent light.

Complimentary Metal Oxide Semiconductor (CMOS) A form of integrated circuit technology that is typically used in low-speed and low-power applications.

Concatenation The technique used in SONET and SDH in which multiple payloads are ganged together to form a super-rate frame capable of transporting payloads greater in size than the basic transmission speed of the system. Thus, an OC-12c provides 622.08 Mbps of total bandwidth, as opposed to an OC-12, which also offers 622.08 Mbps, but in increments of OC-1 (51.84 Mbps).

Connector A device, usually mechanical, used to connect a fiber to a transmit or receive device or to bond two fibers.

Core The central portion of an optical fiber that provides the primary transmission path for an optical signal. It usually has a higher index of refraction than the cladding.

Counter-Rotating Ring A form of transmission system that comprises two rings operating in opposite directions. Typically, one ring serves as the active path while the other serves as the protect or backup path.

Critical Angle The angle at which total internal reflection occurs.

Cross-Phase Modulation (XPM) A problem that occurs in optical fiber that results from the nonlinear index of refraction of the silica in the fiber. Because the index of refraction varies according to the strength of the transmitted signal, some signals interact with each other in destructive ways. Cross-Phase Modulation is considered to be a fiber nonlinearity.

Cutoff Wavelength The wavelength below which single-mode fiber ceases to be single mode.

D

Dark Fiber Optical Fiber that is sometimes leased to a client that is not connected to a transmitter or receiver. In a dark fiber installation, it is the customer's responsibility to terminate the fiber.

Decibel (dB)   A logarithmic measure of the strength of a transmitted signal. Because it is a logarithmic measure, a 20 dB loss would indicate that the received signal is one one-hundredth its original strength.

Dense Wavelength Division Multiplexing (DWDM)   A form of frequency-division multiplexing in which multiple wavelengths of light are transmitted across the same optical fiber. These DWDM systems typically operate in the so-called L-Band (1,625 nm) and have channels that are spaced between 50 and 100 GHz apart. Newly announced products may dramatically reduce this spacing.

Detector   An optical receive device that converts an optical signal into an electrical signal so that it can be handed off to a switch, router, multiplexer, or other electrical transmission device. These devices are usually either NPN or APDs.

Diameter Mismatch Loss   Loss that occurs when the diameter of a light emitter and the diameter of the ingress fiber's core are dramatically different.

Dichroic Filter   A filter that transmits light in a wavelength-specific fashion, reflecting non-selected wavelengths.

Dielectric   A substance that is non-conducting.

Diffraction Grating   A grid of closely spaced lines that are used to selectively direct specific wavelengths of light as required.

Digital   A signal characterized by discrete states. Digital is the opposite of analog.

Digital Hierarchy   In North America, the multiplexing hierarchy that enables 64-Kbps DS-0 signals to be combined to form DS-3 signals for high bit rate transport.

Digital Subscriber Line Access Multiplexer (DSLAM)   The multiplexer in the central office that receives voice and data signals on separate channels, relaying voice to the local switch and data to a router elsewhere in the office.

Diode   A semiconductor device that only allows current to flow in a single direction.

Dispersion   The spreading of a light signal over time that results from modal or chromatic inefficiencies in the fiber.

Dispersion Compensating Fiber (DCF)   A segment of fiber that exhibits the opposite dispersion effect of the fiber to which it is coupled. DCF is used to counteract the dispersion of the other fiber.

Dispersion-Shifted Fiber (DSF)   A form of optical fiber that is designed to exhibit zero dispersion within the C-Band (1,550 nm). DSF does not work well for DWDM because of four-wave mixing problems; non-zero dispersion-shifted fiber is used instead.

Dopant   Substances used to lower the refractive index of the silica used in optical fiber.

DS-0   Digital signal level 0, a 64-Kbps signal.

DS-1   Digital signal level 1, a 1.544-Mbps signal.

DS-2   Digital signal level 2, a 6.312-Mbps signal.

DS-3   Digital signal level 3, a 44.736-Mbps signal.

E

Edge-Emitting Diode   A diode that emits light from the edge of the device rather than the surface, resulting in a more coherent and directed beam of light.

Effective Area   The cross-section of a single-mode fiber that carries the optical signal.

Erbium-Doped Fiber Amplifier (EDFA)   A form of optical amplifier that uses the element erbium to bring about the amplification process. Erbium has the enviable quality that when struck by light operating at 980 nm, it emits photons in the 1,550-nm range, thus providing agnostic amplification for signals operating in the same transmission window.

Evanescent Wave   Light that travels down the inner layer of the cladding instead of down the fiber core.

Eye Pattern   A measure of the degree to which bit errors are occurring in optical transmission systems. The width of the *eyes* (Eye Patterns look like figure eights lying on their sides) indicates the relative bit error rate.

Extrinsic Loss   Loss that occurs at splice points in an optical fiber.

F

Faraday Effect   Sometimes called the magneto-optical effect, the Faraday Effect describes the degree to which some materials can cause the polarization angle of incident light to change when placed within a magnetic field that is parallel to the propagation direction.

Ferrule   A rigid or semi-rigid tube that surrounds optical fibers and protects them.

Fiber Grating   A segment of photosensitive optical fiber that has been treated with ultraviolet light to create a refractive index within the fiber that varies periodically along its length. It operates analogously to a fiber grating and is used to select specific wavelengths of light for transmission.

Fiber-to-the-Curb (FTTC)   A transmission architecture for service delivery in which a fiber is installed in a neighborhood and terminated at a junction box. From there, coaxial cable or twisted pair can be cross-connected from the O-E converter to the customer premises. If coax is used, the system is called Hybrid Fiber Coax (HFC); twisted pair-based systems are called Switched Digital Video (SDV).

Fiber-to-the-Home (FTTH)   Similar to FTTC, except that FTTH extends the optical fiber all the way to the customer premises.

Four-Wave Mixing (FWM)   The nastiest of the so-called fiber nonlinearities. FWM is commonly seen in DWDM systems and occurs when the closely spaced channels mix and generate the equivalent of optical sidebands. The number of these sidebands can be expressed by the equation $1/2(n^3-n^2)$, where n is the number of original channels in the system. Thus, a 16-channel DWDM system will potentially generate 1,920 interfering sidebands!

Frame Relay Bearer Service (FRBS)   In ATM, a service that enables a frame relay frame to be transported across an ATM network.

Fresnel Loss   The loss that occurs at the interface between the head of the fiber and the light source to which it is attached. At air-glass interfaces, the loss usually equates to about 4 percent.

Fused Fiber   A group of fibers that are fused together so that they will remain in alignment. They are often used in one-to-many distribution systems for the propagation of a single signal to multiple destinations. Fused fiber devices play a key role in passive optical networking (PON).

Fusion Splice   A splice made by melting the ends of the fibers together.

G

Generic Flow Control (GFC) In ATM, the first field in the cell header. It is largely unused except when it is overwritten in NNI cells, in which case it becomes additional space for virtual path addressing.

Global Positioning System (GPS) The array of satellites used for radiolocation around the world. In the telephony world, GPS satellites provide an accurate timing signal for synchronizing office equipment.

Graded Index Fiber (GRIN) A type of fiber in which the refractive index changes gradually between the central axis of the fiber and the outer layer, instead of abruptly at the core-cladding interface.

Groom and Fill Similar to add-drop, groom and fill refers to the capability to add (fill) and drop (groom) payload components at intermediate locations along a network path.

H

Header In ATM, the first five bytes of the cell. The header contains information used by the network to route the cell to its ultimate destination. Fields in the cell header include Generic Flow Control, Virtual Path Identifier, Virtual Channel Identifier, Payload Type Identifier, Cell Loss Priority, and Header Error Correction.

Header Error Correction (HEC) In ATM, the header field used to recover from bit errors in the header data.

Hertz (Hz) A measure of cycles per second in transmission systems.

Hybrid Fiber Coax A transmission system architecture in which a fiber feeder penetrates a service area and is then cross-connected to coaxial cable feeders into the customers' premises.

I

Index of refraction A measure of the ratio between the velocity of light in a vacuum and the velocity of the same light in an optical fiber. The refractive index is always greater than one and is denoted n.

Infrared (IR)   The region of the spectrum within which most optical transmission systems operate, found between 700 nm and 0.1 mm.

Injection Laser   A semiconductor laser (synonym).

Intermodulation   A fiber nonlinearity that is similar to four-wave mixing, in which the power-dependent refractive index of the transmission medium allows signals to mix and create destructive sidebands.

Interoperability   In SONET and SDH, the capability of devices from different manufacturers to send and receive information to and from each other successfully.

Intrinsic Loss   Loss that occurs as the result of physical differences in the two fibers being spliced.

Isochronous   A word used in timing systems that means that constant delay occurs across a network.

J

Jacket   The protective outer coating of an optical fiber cable. The jacket may be polyethylene, Kevlar®, or metallic.

Jumper   An optical cable assembly, usually fairly short, that is terminated on both ends with connectors.

L

LAN Emulation (LANE)   In ATM, a service that defines the capability to provide bridging services between LANs across an ATM network.

Large Core Fiber   Fiber that characteristically has a core diameter of 200 microns or more.

Laser   An acronym for Light Amplification by the Stimulated Emission of Radiation. Lasers are used in optical transmission systems because they produce coherent light that is almost purely monochromatic.

Laser Diode (LD)   A diode that produces coherent light when a forward biasing current is applied to it.

Light-Emitting Diode (LED)   A diode that emits incoherent light when a forward bias current is applied to it. LEDs are typically used in shorter distance, lower speed systems.

Lightguide   A term that is used synonymously with optical fiber.

Line Overhead (LOH)   In SONET, the overhead that is used to manage the network regions between multiplexers.

Linewidth   The spectrum of wavelengths that make up an optical signal.

Loose Tube Optical Cable   An optical cable assembly in which the fibers within the cable are loosely contained within tubes inside the sheath of the cable. The fibers are able to move within the tube, thus enabling them to adapt and move without damage as the cable is flexed and stretched.

Loss   The reduction in signal strength that occurs over distance, usually expressed in decibels.

M

M13   A multiplexer that interfaces between DS-1 and DS-3 systems.

Material Dispersion   A dispersion effect caused by the fact that different wavelengths of light travel at different speeds through a medium.

Metasignaling Virtual Channel (MSVC)   In ATM, a signaling channel that is always on. It is used for the establishment of temporary signaling channels as well as channels for voice and data transport.

Microbend   Changes in the physical structure of an optical fiber caused by bending, which can result in light leakage from the fiber.

Mid-span Meet   In SONET and SDH, the term used to describe interoperability. See also *interoperability*.

Modal Dispersion   (See *Multimode Dispersion*)

Mode   A single wave that propagates down a fiber. Multimode fiber enables multiple modes to travel, while single-mode fiber enables only a single mode to be transmitted.

Modulation   The process of changing or *modulating* a carrier wave to cause it to carry information.

Multimode Dispersion   Sometimes referred to as modal dispersion, multimode dispersion is caused by the fact that different modes take different times to move from the ingress point to the egress point of a fiber, thus resulting in modal spreading.

Multimode Fiber   Fiber that has a core diameter of 62.5 microns or greater, wide enough to enable multiple modes of light to be simultaneously transmitted down the fiber.

Multiplexer   A device that has the capability to combine multiple inputs into a single output as a way to reduce the requirement for additional transmission facilities.

Multiprotocol over ATM (MPOA)   In ATM, a service that enables IP packets to be routed across an ATM network.

N

Near-End Crosstalk (NEXT)   The problem that occurs when an optical signal is reflected back toward the input port from one or more output ports. This problem is sometimes referred to as *isolation directivity*.

Non-Dispersion-Shifted Fiber (NDSF)   Fiber that is designed to operate at the low-dispersion second operational window (1,310 nm).

Non-Zero Dispersion-Shifted Fiber (NZDSF)   A form of single-mode fiber that is designed to operate just outside the 1,550 nm window so that fiber nonlinearities, particularly FWM, are minimized.

Numerical Aperture (NA)   A measure of the capability of a fiber to gather light. NA is also a measure of the maximum angle at which a light source can be from the center axis of a fiber in order to collect light.

O

OAM&P   Operations, Administration, Maintenance, and Provisioning, the four key areas in modern network management systems. OAM&P was first coined by the Bell System and continues in widespread use today.

OC-n   Optical Carrier level n, a measure of bandwidth used in SONET systems. OC-1 is 51.84 Mbps; OC-n is *n* times 51.84 Mbps.

Optical Amplifier   A device that amplifies an optical signal without first converting it to an electrical signal.

Optical Carrier Level n (OC-n)   In SONET, the transmission level at which an optical system is operating.

Optical Isolator   A device used to selectively block specific wavelengths of light.

Optical Time Domain Reflectometer (OTDR)   A device used to detect failures in an optical span by measuring the amount of light reflected back from the air-glass interface at the failure point.

Overhead   The part of a transmission stream that the network uses to manage and direct the payload to its destination.

P

Path Overhead   In SONET and SDH, the part of the overhead that is specific to the payload being transported.

Payload   In SONET and SDH, the user data that is being transported.

Payload Type Identifier (PTI)   In ATM, a cell header field that is used to identify network congestion and cell type. The first bit indicates whether the cell was generated by the user or by the network, whereas the second indicates the presence or absence of congestion in user-generated cells, or flow-related Operations, Administration, and Maintenance information in cells generated by the network. The third bit is used for service-specific, higher-layer functions in the user-to-network direction, such as to indicate that a cell is the last in a *series* of cells. From the network to the user, the third bit is used with the second bit to indicate whether the OA&M information refers to segment or end-to-end-related information flow.

Photodetector   A device used to detect an incoming optical signal and convert it to an electrical output.

Photodiode   A semiconductor that converts light to electricity.

Photon   The fundamental unit of light, sometimes referred to as a quantum of electromagnetic energy.

Photonic   The optical equivalent of the term *electronic*.

Planar Waveguide   A waveguide fabricated from a flat material such as a sheet of glass, into which are etched fine lines used to conduct optical signals.

Plenum   The air handling space in buildings found inside walls, under floors, and above ceilings. The plenum spaces are often used as conduits for optical cables.

Plenum Cable   Cable that passes fire retardant tests so that it can legally be used in plenum installations.

Plesiochronous   In timing systems, a term that means "almost synchronized." It refers to the fact that in SONET and SDH systems, payload components frequently derive from different sources, and therefore may have slightly different phase characteristics.

Pointer   In SONET and SDH, a field that is used to indicate the beginning of the transported payload.

Polarization   The process of modifying the direction of the magnetic field within a light wave.

Polarization Mode Dispersion (PMD)   The problem that occurs when light waves with different polarization planes in the same fiber travel at different velocities down the fiber.

Preform   The cylindrical mass of highly pure fused silica from which optical fiber is drawn during the manufacturing process. In the industry, the preform is sometimes referred to as a *gob*.

Pulse Spreading   The widening or spreading out of an optical signal that occurs over distance in a fiber.

Pump Laser   The laser that provides the energy used to excite the dopant in an optical amplifier.

R

Rayleigh Scattering   A scattering effect that occurs in optical fiber as the result of fluctuations in silica density or chemical composition. Metal ions in the fiber often cause Rayleigh Scattering.

Refraction   The change in direction that occurs in a light wave as it passes from one medium into another. The most common example is the bending that is often seen to occur when a stick is inserted into water.

Refractive Index   A measure of the speed at which light travels through a medium, usually expressed as a ration compared to the speed of the same light in a vacuum.

Regenerative Repeater   A device that reconstructs and regenerates a transmitted signal that has been weakened over distance.

Remote Terminal (RT)   In loop carrier systems, the multiplexer located in the field. It communicates with the central office terminal (COT).

S

Scattering   The backsplash or reflection of an optical signal that occurs when it is reflected by small inclusions or particles in the fiber.

SDH   The abbreviation for Synchronous Digital Hierarchy, the European equivalent of SONET.

Section Overhead (SOH)   In SONET systems, the overhead that is used to manage the network regions that occur between repeaters.

Self-Phase Modulation (SPM)   The refractive index of glass is directly related to the power of the transmitted signal. As the power fluctuates, so too does the index of refraction, causing waveform distortion.

Sheath   One of the layers of protective coating in an optical fiber cable.

Signaling Virtual Channel (SVC)   In ATM, a temporary signaling channel used to establish paths for the transport of user traffic.

Single-Mode Fiber (SMF)   The most popular form of fiber today, characterized by the fact that it allows only a single mode of light to propagate down the fiber.

Soliton   A unique waveform that takes advantage of nonlinearities in the fiber medium, the result of which is a signal that suffers essentially no dispersion effects over long distances. Soliton transmission is an area of significant study at the moment because of the promise it holds for long-haul transmission systems.

SONET   Abbreviation for the Synchronous Optical Network, a multiplexing standard that begins at DS3 and provides standards-based multiplexing up to gigabit speeds. SONET is widely used in telephone company long-haul transmission systems and was one of the first widely deployed optical transmission systems.

Source   The emitter of light in an optical transmission system.

Step-Index Fiber   Fiber that exhibits a continuous refractive index in the core, which then steps at the core-cladding interface.

Stimulated Brillouin Scattering (SBS)   A fiber nonlinearity that occurs when a light signal traveling down a fiber interacts with acoustic vibrations in the glass matrix (sometimes called photon-phonon interaction), causing light to be scattered or reflected back toward the source.

Stimulated Raman Scattering (SRS)   A fiber nonlinearity that occurs when power from short-wavelength, high-power channels is bled into longer-wavelength, lower-power channels.

Strength Member   The strand within an optical cable that is used to provide tensile strength to the overall assembly. The member is usually composed of steel, fiberglass, or Aramid yarn.

Surface-Emitting Diode   A semiconductor that emits light from its surface, resulting in a low-power, broad-spectrum emission.

Synchronous   A term that means that both communicating devices derive their synchronization signal from the same source.

Synchronous Transmission Signal Level 1 (STS-1)   In SONET systems, the lowest transmission level in the hierarchy. STS is the electrical equivalent of OC.

T

T-1   In the North American Digital Hierarchy, a 1.544-Mbps signal.

T-3   In the North American Digital Hierarchy, a 44.736-Mbps signal.

Terminal Multiplexer   In SONET and SDH systems, a device that is used to distribute payload to or receive payload from user devices at the end of an optical span.

Tight Buffer Cable   An optical cable in which the fibers are tightly bound by the surrounding material.

Total Internal Reflection   The phenomenon that occurs when light strikes a surface at such an angle that all of the light is reflected back into the transporting medium. In optical fiber, total internal reflection is achieved by keeping the light source and the fiber core oriented along the same axis so that the light that enters the core is reflected back into the core at the core-cladding interface.

Transceiver   A device that incorporates both a transmitter and a receiver in the same housing, thus reducing the need for rack space.

Transponder   A device that incorporates a transmitter, a receiver, and a multiplexer on a single chassis.

V

Vertical Cavity Surface-Emitting Laser (VCSEL)   A small, highly efficient laser that emits light vertically from the surface of the wafer on which it is made.

Virtual Channel (VC)   In ATM, a unidirectional channel between two communicating devices.

Virtual Channel Identifier (VCI)   In ATM, the field that identifies a virtual channel.

Virtual Container   In SDH, the technique used to transport sub-rate payloads.

Virtual Path (VP)   In ATM, a combination of unidirectional virtual channels that make up a bi-directional channel.

Virtual Path Identifier (VPI)   In ATM, the field that identifies a virtual path.

Virtual Tributary (VT)   In SONET, the technique used to transport sub-rate payloads.

Voice/Telephony over ATM (VTOA)   In ATM, a service used to transport telephony signals across an ATM network.

W

Waveguide   A medium that is designed to conduct light within itself over a significant distance, such as optical fiber.

Waveguide Dispersion   A form of chromatic dispersion that occurs when some of the light traveling in the core escapes into the cladding, traveling there at a different speed than the light in the core.

Wavelength   The distance between the same points on two consecutive waves in a chain—for example, from the peak of wave one to the peak of wave two. Wavelength is related to frequency by the equation

$$\lambda = c/f$$

where lambda (λ) is the wavelength, c is the speed of light, and f is the frequency of the transmitted signal.

Wavelength Division Multiplexing (WDM) The process of transmitting multiple wavelengths of light down a fiber.

Window   A region within which optical signals are transmitted at specific wavelengths to take advantage of propagation characteristics that occur there, such as minimum loss or dispersion.

Z

Zero Dispersion Wavelength   The wavelength at which material and waveguide dispersion cancel each other.

BIBLIOGRAPHY

Books

Clarke, Arthur C. *How the World Was One: Beyond the Global Village*. Bantam; New York, 1992.

Evans, Philip and Thomas S. Wurster. *Blown to Bits: How the New Economics of Information Transforms Strategy*. Harvard Business School Press; Boston, Massachusetts, 2000.

Goff, David R. *Fiber Optic Reference Guide: A Practical Guide to the Technology—Second Edition*. Focal Press; Boston, 1999.

Goralski, Walter J. *ADSL and DSL Technologies*. McGraw-Hill; New York, 1998.

Goralski, Walter J. *SONET: A Guide to Synchronous Optical Networks*. McGraw-Hill; New York, 1997.

Goralski, Walter J. and Matthew C. Kolon. *IP Telephony*. McGraw-Hill; New York, 2000.

Hecht, Jeff. *Understanding Fiber Optics—Third Edition*. Prentice-Hall; Upper Saddle River, New Jersey, 1999.

Kartalopoulos, Stamatios. *Understanding SONET / SDH and ATM*. IEEE Press; New York, 1999.

Lanning, Michael J. *Delivering Profitable Value*. Perseus Books; New York, 1998.

Metz, Christopher Y. *IP Switching Protocols and Architectures*. McGraw-Hill; New York, 1999.

Minoli, Daniel and Emma Minoli. *Delivering Voice Over IP Networks*. John Wiley & Sons; New York, 1998.

Rackham, Neil. *Spin Selling*. McGraw-Hill; New York, 1998.

Shepard, Steven. *Telecommunications Convergence: How to Profit from the Convergence of Technologies, Services and Companies*. McGraw-Hill; New York, 2000.

Stern, Thomas E. and Krishna Bala. *Multiwavelength Optical Networks: A Layered Approach*. Addison Wesley Longman; Reading, Massachusetts, 1999.

Tapscott, Don. *Growing Up Digital*. McGraw-Hill; New York, 1998.

Articles

---. *SONET's Still Alive and Kicking.* America's Network; November 15, 2000.

---. *Dense Wavelength Division Multiplexing: from ATG's Communications & Networking Technology Guide Series*; sponsored by Ciena. Available at **www.techguide.com**.

---. *Voice Over Frame Relay: A technical brief by the Frame Relay Forum.* Available at **www.frforum.com/4000/voicetechbrief.html**.

---. *Sphera Optical Networks Takes Advantage of Teledensity in LayerOne's Optical Transport Exchange.* Sphera Optical Networks press release; November 22, 2000.

---. *Optical Switches go Acoustic.* Light Reading; **www.lightreading.com**; June 20, 2000.

---. Notes from an interview with Walter Goralski; October 1999.

---. *A Fever Pitch in Optical Networking.* Business Week Online; April 11, 2000.

---. *Optical Transport Market Booms.* An article from RHK, No date shown. RHK Report on Optical Transport Market 1999.

---. *Cisco Systems Completes the Acquisition of Pirelli Optical Systems.* Cisco press release, February 16, 2000.

---. *A Packet Services Strategy for SONET/SDH Networks.* A white paper from Appian Communications, Inc.

---. *SONET: Who's Ahead in Metro?* Light Reading; March 8, 2001.

---. *SONET: Who's Ahead in Long Haul?* Light Reading; March 6, 2001.

---. *SONET: Who's Ahead?* Light Reading; March 6, 2001.

Abreu, Elinor. *New Fiber in Old Trenches.* The Industry Standard; April 24, 2000.

Allen, Doug. *Dial me a Lambda.* Network Magazine; date unknown.

Allen, Doug. *SONET Rings into Mesh: The Future of Next-Gen Networks.* Network Magazine; April 2001.

Allen, Doug. *Passive Optical Networking Brings DSL to the Masses.* Network Magazine; November 5, 2000.

Aun, Fred and David Hakala. *Bet the House on Optical Networks. That's What Three Networking Giants are Doing. Should You?* ZDNet Networking News; July 13, 2000.

Biagi, Susan. *Fiber Fast and Furious.* Telephony; June 5, 2000.

Branson, Ken. *Local Networks Grow Fiber as Bandwidth Demand Escalates.* X-CHANGE; June 2000.

Cacal, Voltaire D. *The Evolution of Optical Networking.* Notes from a presentation by RHK at the Optical Internetworking Summit; January 20, 2000.

Clavenna, Scott. *The Ultimate Backbone.* Light Reading; October 9, 2000.

Clavenna, Scott. *More Generations of SONET to Come.* Business Communications Review; November 2000.

Crawford, Gregan. *ATM QoS.* Special Supplement to America's Network.

Davis, Christopher C. *Fiber Optic Technology and its Role in the Information Revolution.* **www.ece.umd.edu/~davis/optfib.html** .

Diantina, Pablo. *Latin America Gets More from its Fiber.* FiberSystems International; June/July 2000.

Dominguez, Alex. *Light May Break its Own Speed Limit.* Associated Press; July 19, 2000.

Fairley, Peter. *The Microphotonics Revolution.* Technology Review; July/August 2000.

Fitz, Jonathan G. *Take This Bandwidth and Shove it.* Telephony; June 5, 2000.

Gilder, George and Richard Vigilante. *The Sounds of Silence.* Gilder Technology Report; August 2000, Volume V, Number 8.

Gilder, George and Richard Vigilante. *The Microcosm Strikes at Cisco City.* Gilder Technology Report; September 2000, Volume V, Number 9.

Gonsalves, Chris. *Metromedia Fiber to Buy Web Services Provider SiteSmith.* EWeek; October 11, 2000.

Greenfield, David. *The Optical Revolution.* Network Magazine; February 1, 2000.

HartMayer, Ron. *Components are the Secret of Interactive Networks.* FiberSystems International; June/July 2000.

Heywood, Peter. *Tunable Filters Go Solid State.* Light Reading; November 3, 2000.

Irpan, Thomas and Robert Tsay. *MPLS Technology.* Special Supplement to America's Network.

Isenberg, David. *The Rise of the Stupid Network.* A while paper available at **http//www.isen.com**.

Jander, Mary and Marguerite Reardon. *Corning Hedges its Bets.* Network World Online; September 14, 2000.

Jander, Mary and Marguerite Reardon. *Components Explosion.* Network World Online; December 6, 2000.

Kenward, Michael. *Mirror Magic Ushers in the All-Optical Network.* FiberSystems International; June/July 2000.

Khan, Nisa. *Manufacturers Focus on the 40 Gbit/s Challenge.* FiberSystems International; June/July 2000.

Kim, Philip. *Managing Mega-Networks: What's Happening at the Edge?* Communications News; April 2000.

Kuehn. *Wanted: More Creative Approaches to SONET.* Business Communications Review; December 2000.

Lindstrom, Annie. *Taming the Terrors of the Deep.* A supplement to America's Network.

Lindstrom, Annie. *Unmasking the Fiber Barons, Parts I and II.* America's Network; March 1, 2000.

Llana, Andrés. *Fault Tolerant Network Topologies.* Special Supplement to America's Network.

Lynch, Grahame. *The Next Killer Access Technology: Laser. Is it Reliable? Is it Safe?* America's Network; June 1, 2000.

McGarvey, Joe. *Clear Focus in Optical is on Startups.* Inter@ctive Week; June 26, 2000.

Miller, Elizabeth Starr. *Lucent Catches Metro Fever.* Telephony; June 5, 2000.

Mills, Doyle. *Effective Testing May be More of an Art Than a Science.* Fund at **comnews.com**; date unknown.

Nathan, Dr. Sri. *The Internet's Optical Future.* Communications News; December 1999.

Nolle, Tom. *The RBOCs' Roadmap to Tomorrow's Access Network.* Network Magazine; April 2001.

Nolle, Tom. *The New Order in Networking.* Network Magazine; July 10, 2000.

Raynovich, R. Scott. *Redback Completes IP ASICs.* Light Reading; November 16, 2000.

Reardon, Marguerite. *More Terabit on the Way.* Light Reading; November 22, 2000.

Rigby, Pauline. *Trellis Gets $25M for Holographic Switch.* Light Reading; October 3, 2000.

Rohde, David. *It's Still a Fiber Game for Office Parks.* Network World Online; August 31, 2000.

Rose, Dwane. *DWDM for Everyone.* Telephony; June 5, 2000.

Saudners, Stephen. *Village Unveils "Optical Packet Node."* Light Reading; October 3, 2000.

Saunders, Stephen and Marguerite Reardon. *Organic Lasers: If it's Organic, it's Got to be Good.* Network World Online; August 31, 2000.

Saunders, Stephen and Marguerite Reardon. *Making Optical Switching Crystal Clear.* Network World Online; August 17, 2000.

Schagerlund, Olov. *Roadmap to Optical Networking.* Communications News; November 1999.

Srikanth, Raj and Alex Barros. *Adding Fiber to Your Diet.* Upside; July 2000.

Steinke, Steve. *Fundamentals of Optical Networking.* Network Magazine; June 2000.

Steinke, Steve. *Optical Networking and "Optical Networking."* Network Magazine; June 2000.

Stewart, Alan. *The Great Lambda Wars.* America's Network; May 1, 2000.

Swanson, Bret. *Circling the Fibersphere.* Gilder Technology Report; July 2000, Volume V, Number 7.

Sweeney, Dan. *Mirrors and Smoke: The Optical Challenge.* America's Network; June 1, 2000.

Sweeney, Dan. *Anticipating the Two-Layer Network.* America's Network; February 1, 2001.

Taylor, Steve and Larry Hettick. *Core Migration: It's Mandatory.* Network World Online; September 14, 2000.

Taylor, Steve and Larry Hettick. *Can we Ditch SONET?* Network World Newsletter; March 21, 2001.

Taylor, Steve and Larry Hettick. *Can we Ditch ATM?* Parts 1 and 2. Network World Newsletter; March 26 and 28, 2001.

Thomas, Gordon A., David A. Ackerman, Paul R. Prucnal, S. Lance Cooper. *Physics in the Whirlwind of Optical Communications.* Physics Today; September 2000.

Wirbel, Loring. *Appian Adds Jewel to Optical Ring Protection.* EETimes; August 28, 2000.

Wirbel, Loring. *Optical Pipes Get Thinner, Smarter with Demand.* EETimes; August 30, 2000.

COMMON INDUSTRY ACRONYMS

AAL	ATM Adaptation Layer
AARP	AppleTalk Address Resolution Protocol
ABM	Asynchronous Balanced Mode
ABR	Available Bit Rate
AC	Alternating Current
ACD	Automatic Call Distribution
ACELP	Algebraic Code-Excited Linear Prediction
ACF	Advanced Communication Function
ACK	Acknowledgment
ACM	Address Complete Message
ACSE	Association Control Service Element
ACTLU	Activate Logical Unit
ACTPU	Activate Physical Unit
ADCCP	Advanced Data Communications Control Procedures
ADM	Add/Drop Multiplexer
ADPCM	Adaptive Differential Pulse Code Modulation
ADSL	Asymmetric Digital Subscriber Line
AFI	Authority and Format Identifier
AIN	Advanced Intelligent Network
AIS	Alarm Indication Signal
ALU	Arithmetic Logic Unit
AM	Administrative Module (Lucent 5ESS)
AM	Amplitude Modulation
AMI	Alternate Mark Inversion
AMP	Administrative Module Processor
AMPS	Advanced Mobile Phone System
ANI	Automatic Number Identification (SS7)
ANSI	American National Standards Institute
APD	Avalanche Photodiode
API	Application Programming Interface
APPC	Advanced Program-to-Program Communication

APPN	Advanced Peer-to-Peer Networking
APS	Automatic Protection Switching
ARE	All Routes Explorer (Source Route Bridging)
ARM	Asynchronous Response Mode
ARP	Address Resolution Protocol (IETF)
ARPA	Advanced Research Projects Agency
ARPANET	Advanced Research Projects Agency Network
ARQ	Automatic Repeat Request
ASCII	American Standard Code for Information Interchange
ASI	Alternate Space Inversion
ASIC	Application Specific Integrated Circuit
ASIC	Application-Specific Integrated Circuit
ASK	Amplitude Shift Keying
ASN	Abstract Syntax Notation
ASP	Application Service Provider
AT&T	American Telephone and Telegraph
ATDM	Asynchronous Time Division Multiplexing
ATM	Asynchronous Transfer Mode
ATM	Automatic Teller Machine
ATMF	ATM Forum
AU	Administrative Unit (SDH)
AUG	Administrative Unit Group (SDH)
AWG	American Wire Gauge
B8ZS	Binary 8 Zero Substitution
BANCS	Bell Administrative Network Communications System
BBN	Bolt, Beranak, and Newman
BBS	Bulletin Board Service
Bc	Committed Burst Size
BCC	Blocked Calls Cleared
BCC	Block Check Character
BCD	Blocked Calls Delayed
BCDIC	Binary Coded Decimal Interchange Code

Be	Excess Burst Size
BECN	Backward Explicit Congestion Notification
BER	Bit Error Rate
BERT	Bit Error Rate Test
BGP	Border Gateway Protocol (IETF)
BIB	Backward Indicator Bit (SS7)
B-ICI	Broadband Intercarrier Interface
BIOS	Basic Input/Output System
BIP	Bit Interleaved Parity
B-ISDN	Broadband Integrated Services Digital Network
BISYNC	Binary Synchronous Communications Protocol
BITNET	Because It's Time Network
BITS	Building Integrated Timing Supply
BLSR	Bidirectional Line Switched Ring
BLSR	Bidirectional Line Switched Ring
BOC	Bell Operating Company
BPRZ	Bipolar Return to Zero
Bps	Bits per Second
BRI	Basic Rate Interface
BRITE	Basic Rate Interface Transmission Equipment
BSC	Binary Synchronous Communications
BSN	Backward Sequence Number (SS7)
BSRF	Bell System Reference Frequency
BTAM	Basic Telecommunications Access Method
BUS	Broadcast Unknown Server
C/R	Command/Response
CAD	Computer-Aided Design
CAE	Computer-Aided Engineering
CAM	Computer-Aided Manufacturing
CAP	Carrierless Amplitude/Phase modulation
CAP	Competitive Access Provider
CARICOM	Caribbean Community and Common Market

CASE	Common Application Service Element
CASE	Computer-Aided Software Engineering
CAT	Computer-Aided Tomography
CATIA	Computer-Assisted Three-dimensional Interactive Application
CATV	Community Antenna Television
CBEMA	Computer and Business Equipment Manufacturers Association
CBR	Constant Bit Rate
CBT	Computer-Based Training
CC	Cluster Controller
CCIR	International Radio Consultative Committee
CCIS	Common Channel Interoffice Signaling
CCITT	International Telegraph and Telephone Consultative Committee
CCS	Common Channel Signaling
CCS	Hundred Call Seconds per Hour
CD	Collision Detection
CD	Compact Disc
CDC	Control Data Corporation
CDMA	Code Division Multiple Access
CDPD	Cellular Digital Packet Data
CD-ROM	Compact Disc-Read Only Memory
CDVT	Cell Delay Variation Tolerance
CEI	Comparably Efficient Interconnection
CEPT	Conference of European Postal and Telecommunications Administrations
CERN	European Council for Nuclear Research
CERT	Computer Emergency Response Team
CES	Circuit Emulation Service
CEV	Controlled Environmental Vault
CGI	Common Gateway Interface (Internet)
CHAP	Challenge Handshake Authentication Protocol

CICS	Customer Information Control System
CICS/VS	Customer Information Control System/Virtual Storage
CIDR	Classless Interdomain Routing (IETF)
CIF	Cells In Frames
CIR	Committed Information Rate
CISC	Complex Instruction Set Computer
CIX	Commercial Internet Exchange
CLASS	Custom Local Area Signaling Services (Bellcore)
CLEC	Competitive Local Exchange Carrier
CLLM	Consolidated Link Layer Management
CLNP	Connectionless Network Protocol
CLNS	Connectionless Network Service
CLP	Cell Loss Priority
CM	Communications Module (Lucent 5ESS)
CMIP	Common Management Information Protocol
CMISE	Common Management Information Service Element
CMOL	CMIP Over LLC
CMOS	Complementary Metal Oxide Semiconductor
CMOT	CMIP Over TCP/IP
CMP	Communications Module Processor
CNE	Certified NetWare Engineer
CNM	Customer Network Management
CNR	Carrier-to-Noise Ratio
CO	Central Office
CoCOM	Coordinating Committee on Export Controls
CODEC	Coder-Decoder
COMC	Communications Controller
CONS	Connection-Oriented Network Service
CORBA	Common Object Request Brokered Architecture
COS	Class of Service (APPN)
COS	Corporation for Open Systems
CPE	Customer Premises Equipment

CPU	Central Processing Unit
CRC	Cyclic Redundancy Check
CRT	Cathode Ray Tube
CRV	Call Reference Value
CS	Convergence Sublayer
CSA	Carrier Serving Area
CSMA	Carrier Sense Multiple Access
CSMA/CA	Carrier Sense Multiple Access with Collision Avoidance
CSMA/CD	Carrier Sense Multiple Access with Collision Detection
CSU	Channel Service Unit
CTI	Computer Telephony Integration
CTIA	Cellular Telecommunications Industry Association
CTS	Clear To Send
CU	Control Unit
CVSD	Continuously Variable Slope Delta modulation
CWDM	Coarse Wavelength Division Multiplexing
D/A	Digital-to-Analog
DA	Destination Address
DAC	Dual Attachment Concentrator (FDDI)
DACS	Digital Access and Cross-connect System
DARPA	Defense Advanced Research Projects Agency
DAS	Dual Attachment Station (FDDI)
DASD	Direct Access Storage Device
DB	Decibel
DBS	Direct Broadcast Satellite
DC	Direct Current
DCC	Data Communications Channel (SONET)
DCE	Data Circuit-terminating Equipment
DCN	Data Communications Network
DCS	Digital Cross-connect System
DCT	Discrete Cosine Transform
DDCMP	Digital Data Communications Management Protocol (DNA)

DDD	Direct Distance Dialing
DDP	Datagram Delivery Protocol
DDS	DATAPHONE Digital Service (Sometimes Digital Data Service)
DDS	Digital Data Service
DE	Discard Eligibility (LAPF)
DECT	Digital European Cordless Telephone
DES	Data Encryption Standard (NIST)
DID	Direct Inward Dialing
DIP	Dual Inline Package
DLC	Digital Loop Carrier
DLCI	Data Link Connection Identifier
DLE	Data Link Escape
DLSw	Data Link Switching
DM	Delta Modulation
DM	Disconnected Mode
DMA	Direct Memory Access (computers)
DMAC	Direct Memory Access Control
DME	Distributed Management Environment
DMS	Digital Multiplex Switch
DNA	Digital Network Architecture
DNIC	Data Network Identification Code (X.121)
DNIS	Dialed Number Identification Service
DNS	Domain Name System (IETF)
DOD	Direct Outward Dialing
DOD	Department of Defense
DOJ	Department of Justice
DOV	Data Over Voice
DPSK	Differential Phase Shift Keying
DQDB	Distributed Queue Dual Bus
DRAM	Dynamic Random Access Memory
DSAP	Destination Service Access Point
DSF	Dispersion-Shifted Fiber

DSI	Digital Speech Interpolation
DSL	Digital Subscriber Line
DSLAM	Digital Subscriber Line Access Multiplexer
DSP	Digital Signal Processing
DSR	Data Set Ready
DSS	Digital Satellite System
DSS	Digital Subscriber Signaling System
DSU	Data Service Unit
DTE	Data Terminal Equipment
DTMF	Dual Tone Multifrequency
DTR	Data Terminal Ready
DVRN	Dense Virtual Routed Networking (Crescent)
DWDM	Dense Wavelength Division Multiplexing
DXI	Data Exchange Interface
E/O	Electrical-to-Optical
EBCDIC	Extended Binary Coded Decimal Interchange Code
ECMA	European Computer Manufacturer Association
ECN	Explicit Congestion Notification
ECSA	Exchange Carriers Standards Association
EDFA	Erbium-Doped Fiber Amplifier
EDI	Electronic Data Interchange
EDIBANX	EDI Bank Alliance Network Exchange
EDIFACT	Electronic Data Interchange For Administration, Commerce, and Trade (ANSI)
EFCI	Explicit Forward Congestion Indicator
EFTA	European Free Trade Association
EGP	Exterior Gateway Protocol (IETF)
EIA	Electronics Industry Association
EIGRP	Enhanced Interior Gateway Routing Protocol
EIR	Excess Information Rate
EMBARC	Electronic Mail Broadcast to a Roaming Computer
EMI	Electromagnetic Interference
EMS	Element Management System

EN	End Node
ENIAC	Electronic Numerical Integrator and Computer
EO	End Office
EOC	Embedded Operations Channel (SONET)
EOT	End of Transmission (BISYNC)
EPROM	Erasable Programmable Read Only Memory
ESCON	Enterprise System Connection (IBM)
ESF	Extended Superframe Format
ESP	Enhanced Service Provider
ESS	Electronic Switching System
ETSI	European Telecommunications Standards Institute
ETX	End of Text (BISYNC)
EWOS	European Workshop for Open Systems
FACTR	Fujitsu Access and Transport System
FAQ	Frequently Asked Questions
FAT	File Allocation Table
FCS	Frame Check Sequence
FDD	Frequency Division Duplex
FDDI	Fiber Distributed Data Interface
FDM	Frequency Division Multiplexing
FDMA	Frequency Division Multiple Access
FDX	Full-Duplex
FEBE	Far End Block Error (SONET)
FEC	Forward Error Correction
FEC	Forward Equivalence Class
FECN	Forward Explicit Congestion Notification
FEP	Front-End Processor
FERF	Far End Receive Failure (SONET)
FET	Field Effect Transistor
FHSS	Frequency Hopping Spread Spectrum
FIB	Forward Indicator Bit (SS7)
FIFO	First In First Out

FITL	Fiber In The Loop
FLAG	Fiber Ling Across the Globe
FM	Frequency Modulation
FPGA	Field Programmable Gate Array
FR	Frame Relay
FRAD	Frame Relay Access Device
FRBS	Frame Relay Bearer Service
FSK	Frequency Shift Keying
FSN	Forward Sequence Number (SS7)
FTAM	File Transfer, Access, and Management
FTP	File Transfer Protocol (IETF)
FTTC	Fiber to the Curb
FTTH	Fiber to the Home
FUNI	Frame User-to-Network Interface
FWM	Four Wave Mixing
GATT	General Agreement on Tariffs and Trade
GbE	Gigabit Ethernet
Gbps	Gigabits per Second (Billion bits per second)
GDMO	Guidelines for the Development of Managed Objects
GEOS	Geosynchronous Earth Orbit Satellites
GFC	Generic Flow Control (ATM)
GFI	General Format Identifier (X.25)
GOSIP	Government Open Systems Interconnection Profile
GPS	Global Positioning System
GRIN	Graded Index (fiber)
GSM	Global System for Mobile Communications
GUI	Graphical User Interface
HDB3	High Density, Bipolar 3 (E-Carrier)
HDLC	High-level Data Link Control
HDSL	High-bit-rate Digital Subscriber Line
HDTV	High Definition Television
HDX	Half-Duplex

HEC	Header Error Control (ATM)
HFC	Hybrid Fiber/Coax
HFS	Hierarchical File Storage
HLR	Home Location Register
HSSI	High-Speed Serial Interface (ANSI)
HTML	Hypertext Markup Language
HTTP	Hypertext Transfer Protocol (IETF)
HTU	HDSL Transmission Unit
I	Intrapictures
IAB	Internet Architecture Board (formerly Internet Activities Board)
IACS	Integrated Access and Cross-connect System
IAD	Integrated Access Device
IAM	Initial Address Message (SS7)
IANA	Internet Address Naming Authority
ICMP	Internet Control Message Protocol (IETF)
IDP	Internet Datagram Protocol
IEC	Interexchange Carrier (also IXC)
IEC	International Electrotechnical Commission
IEEE	Institute of Electrical and Electronics Engineers
IETF	Internet Engineering Task Force
IFRB	International Frequency Registration Board
IGP	Interior Gateway Protocol (IETF)
IGRP	Interior Gateway Routing Protocol
ILEC	Incumbent Local ExchangeCarrier
IML	Initial Microcode Load
IMP	Interface Message Processor (ARPANET)
IMS	Information Management System
InARP	Inverse Address Resolution Protocol (IETF)
InATMARP	Inverse ATMARP
INMARSAT	International Maritime Satellite Organization
INP	Internet Nodal Processor
InterNIC	Internet Network Information Center

IP	Internet Protocol (IETF)
IPX	Internetwork Packet Exchange (NetWare)
ISDN	Integrated Services Digital Network
ISO	International Organization for Standardization
ISOC	Internet Society
ISP	Internet Service Provider
ISUP	ISDN User Part (SS7)
IT	Information Technology
ITU	International Telecommunication Union
ITU-R	International Telecommunication Union-Radio Communication Sector
IVD	Inside Vapor Deposition
IVR	Interactive Voice Response
IXC	Interexchange Carrier
JEPI	Joint Electronic Paynets Initiative
JES	Job Entry System
JIT	Just in Time
JPEG	Joint Photographic Experts Group
KB	Kilobytes
Kbps	Kilobits per Second (Thousand Bits per Second)
KLTN	Potassium Lithium Tantalate Niobate
LAN	Local Area Network
LANE	LAN Emulation
LAP	Link Access Procedure (X.25)
LAPB	Link Access Procedure Balanced (X.25)
LAPD	Link Access Procedure for the D-Channel
LAPF	Link Access Procedure to Frame Mode Bearer Services
LAPF-Core	Core Aspects of the Link Access Procedure to Frame Mode Bearer Services
LAPM	Link Access Procedure for Modems
LAPX	Link Access Procedure half-duplex
LASER	Light Amplification by the Stimulated Emission of Radiation

LATA	Local Access and Transport Area
LCD	Liquid Crystal Display
LCGN	Logical Channel Group Number
LCM	Line Concentrator Module
LCN	Local Communications Network
LD	Laser Diode
LDAP	Lightweight Directory Access Protocol (X.500)
LEAF®	Large Effective Area Fiber® (Corning product)
LEC	Local Exchange Carrier
LED	Light Emitting Diode
LENS	Lightwave Efficient Network Solution (Centerpoint)
LEOS	Low Earth Orbit Satellites
LI	Length Indicator
LIDB	Line Information Database
LIFO	Last In First Out
LIS	Logical IP Subnet
LLC	Logical Link Control
LMDS	Local Multipoint Distribution System
LMI	Local Management Interface
LMOS	Loop Maintenance Operations System
LORAN	Long-range Radio Navigation
LPC	Linear Predictive Coding
LPP	Lightweight Presentation Protocol
LRC	Longitudinal Redundancy Check (BISYNC)
LS	Link State
LSI	Large Scale Integration
LSP	Label Switched Path
LU	Line Unit
LU	Logical Unit (SNA)
MAC	Media Access Control
MAN	Metropolitan Area Network
MAP	Manufacturing Automation Protocol

MAU	Medium Attachment Unit (Ethernet)
MAU	Multistation Access Unit (Token Ring)
MB	Megabytes
MBA™	Metro Business Access™ (Ocular)
Mbps	Megabits per Second (Million bits per second)
MD	Message Digest (MD2, MD4, MD5) (IETF)
MDF	Main Distribution Frame
MEMS	Micro Electrical Mechanical System
MF	Multifrequency
MFJ	Modified Final Judgment
MHS	Message Handling System (X.400)
MIB	Management Information Base
MIC	Medium Interface Connector (FDDI)
MIME	Multipurpose Internet Mail Extensions (IETF)
MIPS	Millions of Instructions Per Second
MIS	Management Information Systems
MITI	Ministry of International Trade and Industry (Japan)
ML-PPP	Multilink Point-to-Point Protocol
MMDS	Multichannel, Multipoint Distribution System
MMF	Multimode Fiber
MNP	Microcom Networking Protocol
MP	Multilink PPP
MPEG	Motion Picture Experts Group
MPLS	Multiprotocol Label Switching
MPOA	Multiprotocol Over ATM
MRI	Magnetic Resonance Imaging
MSB	Most Significant Bit
MSC	Mobile Switching Center
MSO	Mobile Switching Office
MSVC	Meta-signaling Virtual Channel
MTA	Major Trading Area
MTBF	Mean Time Between Failure

MTP	Message Transfer Part (SS7)
MTTR	Mean Time to Repair
MTU	Maximum Transmission Unit
MVS	Multiple Virtual Storage
NAFTA	North American Free Trade Agreement
NAK	Negative Acknowledgment (BISYNC, DDCMP)
NAP	Network Access Point (Internet)
NARUC	National Association of Regulatory Utility Commissioners
NASA	National Aeronautics and Space Administration
NASDAQ	National Association of Securities Dealers Automated Quotations
NATA	North American Telecommunications Association
NATO	North Atlantic Treaty Organization
NAU	Network Accessible Unit
NCP	Network Control Program
NCSA	National Center for Supercomputer Applications
NCTA	National Cable Television Association
NDIS	Network Driver Interface Specifications
NDSF	Non-Dispersion-Shifted Fiber
NetBEUI	NetBIOS Extended User Interface
NetBIOS	Network Basic Input/Output System
NFS	Network File System (Sun)
NIC	Network Interface Card
NII	National Information Infrastructure
NIST	National Institute of Standards and Technology (formerly NBS)
NIU	Network Interface Unit
NLPID	Network Layer Protocol Identifier
NLSP	NetWare Link Services Protocol
NM	Network Module
Nm	Nanometer
NMC	Network Management Center
NMS	Network Management System

NMT	Nordic Mobile Telephone
NMVT	Network Management Vector Transport protocol
NNI	Network Node Interface
NNI	Network-to-Network Interface
NOC	Network Operations Center
NOCC	Network Operations Control Center
NOS	Network Operating System
NPA	Numbering Plan Area
NREN	National Research and Education Network
NRZ	Non-Return to Zero
NRZI	Non-Return to Zero Inverted
NSA	National Security Agency
NSAP	Network Service Access Point
NSAPA	Network Service Access Point Address
NSF	National Science Foundation
NTSC	National Television SystemsCommittee
NTT	Nippon Telephone and Telegraph
NVOD	Near Video on Demand
NZDSF	Non-Zero Dispersion-Shifted Fiber
OADM	Optical Add-Drop Multiplexer
OAM	Operations, Administration, and Maintenance
OAM&P	Operations, Administration, Maintenance, and Provisioning
OAN	Optical Area Network
OC	Optical Carrier
OEM	Original Equipment Manafacturer
O-E-O	Optical-Electrical-Optical
OLS	Optical Line System (Lucent)
OMAP	Operations, Maintenance, and Administration Part (SS7)
ONA	Open Network Architecture
ONU	Optical Network Unit
OOF	Out of Frame
OS	Operating System

OSF	Open Software Foundation
OSI	Open Systems Interconnection (ISO, ITU-T)
OSI RM	Open Systems Interconnection Reference Model
OSPF	Open Shortest Path First (IETF)
OSS	Operation Support Systems
OTDM	Optical Time Division Multiplexing
OTDR	Optical Time-Domain Reflectometer
OUI	Organizationally Unique Identifier (SNAP)
OVD	Outside Vapor Deposition
P/F	Poll/Final (HDLC)
PAD	Packet Assembler/Disassembler (X.25)
PAL	Phase Alternate Line
PAM	Pulse Amplitude Modulation
PANS	Pretty Amazing New Stuff
PBX	Private Branch Exchange
PCI	Pulse Code Modulation
PCMCIA	Personal Computer Memory Card International Association
PCN	Personal Communications Network
PCS	Personal Communications Services
PDA	Personal Digital Assistant
PDH	Plesiochronous Digital Hierarchy
PDU	Protocol Data Unit
PIN	Positive-Intrinsic-Negative
PING	Packet Internet Groper (TCP/IP)
PLCP	Physical Layer Convergence Protocol
PLP	Packet Layer Protocol (X.25)
PM	Phase Modulation
PMD	Physical Medium Dependent (FDDI)
PNNI	Private Network Node Interface (ATM)
PON	Passive Optical Networking
POP	Point of Presence
POSIT	Profiles for Open Systems Interworking Technologies

POSIX	Portable Operating System Interface for UNIX
POTS	Plain Old Telephone Service
PPP	Point-to-Point Protocol (IETF)
PRC	Primary Reference Clock
PRI	Primary Rate Interface
PROFS	Professional Office System
PROM	Programmable Read Only Memory
PSDN	Packet Switched Data Network
PSK	Phase Shift Keying
PSPDN	Packet Switched Public Data Network
PSTN	Public Switched Telephone Network
PTI	Payload Type Identifier (ATM)
PTT	Post, Telephone, and Telegraph
PU	Physical Unit (SNA)
PUC	Public Utility Commission
PVC	Permanent Virtual Circuit
QAM	Quadrature Amplitude Modulation
Q-bit	Qualified data bit (X.25)
QLLC	Qualified Logical Link Control (SNA)
QoS	Quality of Service
QPSK	Quadrature Phase Shift Keying
QPSX	Queued Packet Synchronous Exchange
R&D	Research & Development
RADSL	Rate Adaptive Digital Subscriber Line
RAID	Redundant Array of Inexpensive Disks
RAM	Random Access Memory
RARP	Reverse Address Resolution Protocol (IETF)
RAS	Remote Access Server
RBOC	Regional Bell Operating Company
RF	Radio Frequency
RFC	Request For Comments (IETF)
RFH	Remote Frame Handler (ISDN)

RFI	Radio Frequency Interference
RFP	Request For Proposal
RHC	Regional Holding Company
RHK	Ryan, Hankin and Kent (Consultancy)
RIP	Routing Information Protocol (IETF)
RISC	Reduced Instruction Set Computer
RJE	Remote Job Entry
RNR	Receive Not Ready (HDLC)
ROM	Read-Only Memory
ROSE	Remote Operation Service Element
RPC	Remote Procedure Call
RR	Receive Ready (HDLC)
RTS	Request To Send (EIA-232-E)
S/DMS	SONET/Digital Multiplex System
S/N	Signal-to-Noise Ratio
SAA	Systems Application Architecture (IBM)
SAAL	Signaling ATM Adaptation Layer (ATM)
SABM	Set Asynchronous Balanced Mode (HDLC)
SABME	Set Asynchronous Balanced Mode Extended (HDLC)
SAC	Single Attachment Concentrator (FDDI)
SAN	Storage Area Network
SAP	Service Access Point (generic)
SAPI	Service Access Point Identifier (LAPD)
SAR	Segmentation and Reassembly (ATM)
SAS	Single Attachment Station (FDDI)
SASE	Specific Applications Service Element (subset of CASE, Application Layer)
SATAN	System Administrator Tool for Analyzing Networks
SBS	Stimulated Brillouin Scattering
SCCP	Signaling Connection Control Point (SS7)
SCP	Service Control Point (SS7)
SCREAM™	Scalable Control of a Rearrangeable Extensible Array of Mirrors (Calient)

SCSI	Small Computer Systems Interface
SCTE	Serial Clock Transmit External (EIA-232-E)
SDH	Synchronous Digital Hierarchy (ITU-T)
SDLC	Synchronous Data Link Control (IBM)
SDS	Scientific Data Systems
SECAM	Sequential Color with Memory
SF	Superframe Format (T-1)
SGML	Standard Generalized Markup Language
SGMP	Simple Gateway Management Protocol (IETF)
S-HTTP	Secure HTTP (IETF)
SIF	Signaling Information Field
SIG	Special Interest Group
SIO	Service Information Octet
SIR	Sustained Information Rate (SMDS)
SLA	Service Level Agreement
SLIP	Serial Line Interface Protocol (IETF)
SM	Switching Module
SMAP	System Management Application Part
SMDS	Switched Multimegabit Data Service
SMF	Single Mode Fiber
SMP	Simple Management Protocol
SMP	Switching Module Processor
SMR	Specialized Mobile Radio
SMS	Standard Management System (SS7)
SMTP	Simple Mail Transfer Protocol (IETF)
SNA	Systems Network Architecture (IBM)
SNAP	Subnetwork Access Protocol
SNI	Subscriber Network Interface (SMDS)
SNMP	Simple Network Management Protocol (IETF)
SNP	Sequence Number Protection
SONET	Synchronous Optical Network
SPAG	Standards Promotion and Application Group

SPARC	Scalable Performance Architecture
SPE	Synchronous Payload Envelope (SONET)
SPID	Service Profile Identifier (ISDN)
SPM	Self Phase Modulation
SPOC	Single Point of Contact
SPX	Sequenced Packet Exchange (NetWare)
SQL	Structured Query Language
SRB	Source Route Bridging
SRS	Stimulated Raman Scattering
SRT	Source Routing Transparent
SS7	Signaling System 7
SSL	Secure Socket Layer (IETF)
SSP	Service Switching Point (SS7)
SST	Spread Spectrum Transmission
STDM	Statistical Time Division Multiplexing
STM	Synchronous Transfer Mode
STM	Synchronous Transport Module (SDH)
STP	Signal Transfer Point (SS7)
STS	Synchronous Transport Signal (SONET)
STX	Start of Text (BISYNC)
SVC	Signaling Virtual Channel (ATM)
SVC	Switched Virtual Circuit
SXS	Step-by-Step Switching
SYN	Synchronization
SYNTRAN	Synchronous Transmission
TA	Terminal Adapter (ISDN)
TAG	Technical Advisory Group
TASI	Time Assigned Speech Interpolation
TAXI	Transparent Asynchronous Transmitter/Receiver Interface (Physical Layer)
TCAP	Transaction Capabilities Application Part (SS7)
TCM	Time Compression Multiplexing
TCM	Trellis Coding Modulation

TCP	Transmission Control Protocol (IETF)
TDD	Time Division Duplexing
TDM	Time Division Multiplexing
TDM	Time Division Multiplexing
TDMA	Time Division Multiple Access
TDR	Time Domain Reflectometer
TE1	Terminal Equipment type 1 (ISDN capable)
TE2	Terminal Equipment type 2 (non-ISDN capable)
TEI	Terminal Endpoint Identifier (LAPD)
TELRIC	Total Element Long-Run Incremental Cost
TIA	Telecommunications Industry Association
TIRKS	Trunk Integrated Record Keeping System
TL1	Transaction Language 1
TM	Terminal Multiplexer
TMN	Telecommunications Management Network
TMS	Time-Multiplexed Switch
TOH	Transport Overhead (SONET)
TOP	Technical and Office Protocol
TOS	Type of Service (IP)
TP	Twisted Pair
TR	Token Ring
TRA	Traffic Routing Administration
TSI	Time Slot Interchange
TSLRIC	Total Service Long-Run Incremental Cost
TSO	Terminating Screening Office
TSO	Time-Sharing Option (IBM)
TSR	Terminate and Stay Resident
TSS	Telecommunication Standardization Sector (ITU-T)
TST	Time-Space-Time Switching
TSTS	Time-Space-Time-Space Switching
TTL	Time to Live
TU	Tributary Unit (SDH)

TUG	Tributary Unit Group (SDH)
TUP	Telephone User Part (SS7)
UA	Unnumbered Acknowledgment (HDLC)
UART	Universal Asynchronous Receiver Transmitter
UBR	Unspecified Bit Rate (ATM)
UDI	Unrestricted Digital Information (ISDN)
UDP	User Datagram Protocol (IETF)
UHF	Ultra High Frequency
UI	Unnumbered Information (HDLC)
UNI	User-to-Network Interface (ATM, FR)
UNIT™	Unified Network Interface Technology™ (Ocular)
UNMA	Unified Network Management Architecture
UPS	Uninterruptable Power Supply
UPSR	Unidirectional Path Switched Ring
UPT	Universal Personal Telecommunications
URL	Uniform Resource Locator
USART	Universal Synchronous Asynchronous Receiver Transmitter
UTC	Coordinated Universal Time
UTP	Unshielded Twisted Pair (Physical Layer)
UUCP	UNIX-UNIX Copy
VAN	Value-Added Network
VAX	Virtual Address Extension (DEC)
vBNS	Very High Speed Backbone Network Service
VBR	Variable Bit Rate (ATM)
VBR-NRT	Variable Bit Rate-Non-Real-Time (ATM)
VBR-RT	Variable Bit Rate-Real-Time (ATM)
VC	Virtual Channel (ATM)
VC	Virtual Circuit (PSN)
VC	Virtual Container (SDH)
VCC	Virtual Channel Connection (ATM)
VCI	Virtual Channel Identifier (ATM)
VCI	Virtual Channel Identifier (ATM)

VCSEL	Vertical Cavity Surface Emitting Laser
VDSL	Very High-speed Digital Subscriber Line
VDSL	Very High bit rate Digital Subscriber Line
VERONICA	Very Easy Rodent-Oriented Netwide Index to Computerized Archives (Internet)
VGA	Variable Graphics Array
VHF	Very High Frequency
VHS	Video Home System
VINES	Virtual Networking System (Banyan)
VIP	VINES Internet Protocol
VLF	Very Low Frequency
VLR	Visitor Location Register (Wireless/GSM)
VLSI	Very Large Scale Integration
VM	Virtual Machine (IBM)
VM	Virtual Memory
VMS	Virtual Memory System (DEC)
VOD	Video-on-Demand
VP	Virtual Path
VPC	Virtual Path Connection
VPI	Virtual Path Identifier
VPN	Virtual Private Network
VPN	Virtual Private Network
VR	Virtual Reality
VSAT	Very Small Aperture Terminal
VSB	Vestigial Sideband
VSELP	Vector-Sum Excited Linear Prediction
VT	Virtual Tributary
VTAM	Virtual Telecommunications Access Method (SNA)
VTOA	Voice and Telephony over ATM
VTP	Virtual Terminal Protocol (ISO)
WACK	Wait Acknowledgment (BISYNC)
WACS	Wireless Access Communications System
WAIS	Wide Area Information Server (IETF)

WAN	Wide Area Network
WARC	World Administrative Radio Conference
WATS	Wide Area Telecommunications Service
WDM	Wavelength Division Multiplexing
WIN	Wireless In-building Network
WTO	World Trade Organization
WWW	World Wide Web (IETF)
WYSIWYG	What You See Is What You Get
xDSL	x-Type Digital Subscriber Line
XID	Exchange Identification (HDLC)
XNS	Xerox Network Systems
XPM	Cross Phase Modulation
ZBTSI	Zero Byte Time Slot Interchange
ZCS	Zero Code Suppression

ONLINE RESOURCES

Agere Systems	www.agere.com
Agilent	www.agilent.com
Agility	www.agility.com
AirFiber	www.airfiber.com
Alcatel	www.alcatel.com
American National Standards Institute (ANSI)	www.ansi.org
Appian Communications	www.appiancommunications.com
Astral Point	www.astralpoint.com
ATG Advanced TelCom Group	www2.callatg.com
Boston Optical	www.bostonoptical.com
Broadwing	www.broadwing.com
Chromatis	www.chromatis.com
Cisco	www.cisco.com
Colo.com	www.colo.com
Corning, Inc.	www.corningfiber.com
Corvis	www.corvis.com
Crescent Networks	www.crescent.com
Electronic Industries Association (EIA)	www.eia.org
European Telecommunications Standards Institute (ETSI)	www.etsi.org
Extreme Networks	www.extremenetworks.com
Fiber Optics Online	www.fiberopticsonline.com
Fibercore	www.fibercore.com
FPN Magazine	www.fpnmag.com
Fujikura	www.fujikura.com
Furukawa Electric Co.	www.furukawa.co.jp/english
Institute of Electrical and Electronics Engineers (IEEE)	www.ieee.org
International Telecommunications Union (ITU)	www.itu.int
JDS Uniphase	www.jdsuniphase.com
Light Reading	www.lightreading.com
Lightwave Microsystems	www.lightwavemicrosystems.com
Lucent	www.lucent.com
Mayan Networks	www.mayannetworks.com
Nanovation Technologies	www.nanovation.com

NEC	`www.nec-global.com`
Network World	`www.nwfusion.com/topics/optical.html`
Nortel	`www.nortelnetworks.com`
Novalux	`www.novalux.com`
Ocular Networks	`www.ocularnetworks.com`
Optical Domain Service Interconnect (ODSI)	`www.odsi-coalition.com`
Optical Internetworking Forum	`www.oiforum.com`
Optical Society of America	`www.osa.org`
Optical Solutions	`www.opticalsolutions.com`
Optronics	`www.optronics.com`
Pluris	`www.pluris.com`
Polatis	`www.polatis.com`
Qwest	`www.qwest.com`
Redback Networks	`www.redbacknetworks.com`
SONET Interoperability Forum (SIF)	`www.atis.org/atis/sif`
Southampton Photonics	`www.southamptonphotonics.com`
Sphera Networks	`www.spheranetworks.com`
Startech	`www.startech.com/fiberoptics/`
Sycamore	`www.sycamore.com`
Tacjion	`www.tachion.com`
Telcordia	`www.telcordia.com`
Telecommunications Industry Association	`www.tiaonline.org`
Terabeam	`www.terabeam.com`
Ultra Fast Optical Systems	`www.ufos.com`
Yipes	`www.yipes.com`

TELCORDIA DOCUMENTATION

MC-APD-CS-020	The Evolution of SONET Operations	Apr 2000
MC-APD-CS-021	TMN Planning and Deployment Support for SONET and ATM Networks	Apr 2000
MC-APD-CS-033	Evolving a Profit-Building Provider Network	Apr 2000
MC-APD-CS-039	Transport Product Concept Evaluation	May 2000
MC-APD-CS-040	SONET Product Gap Analysis	May 2000
MC-APD-CS-042	Analysis of Mesh-Based and Ring-Based Transport Network Architectures	May 2000
MC-APD-CS-043	Analysis of the North American SONET Market	May 2000
MC-APD-CS-044	SONET Product Portfolio Analysis	May 2000
MC-TAI-SO-018	SONET Testing, Analysis, and Deployment Support for Integrated Communications Providers	Jan 2001
MC-TAI-SO-028	Transport (SONET/SDH) Product Analysis Services for Suppliers	Jan 2001

Family of Requirements

FR-SONET-17	Broadband and Transport Network Generic Requirements: SONET and ATM Transport Technologies	Apr 2001
FR-SONET-17-CD-1USER	Broadband and Transport Network Generic Requirements: SONET and ATM Transport Technologies	Apr 2001
FR-SONET-17-CD-2-5USER	Broadband and Transport Network Generic Requirements: SONET and ATM Transport Technologies	Apr 2001
FR-SONET-17-CD-21-100USER	Broadband and Transport Network Generic Requirements: SONET and ATM Transport Technologies	Apr 2001

| FR-SONET-17-CD-6-20USER | Broadband and Transport Network Generic Requirements: SONET and ATM Transport Technologies | Apr 2001 |

Generic Requirements

GR-1042	Generic requirements for operations interfaces using OSI tools—Information model overview: synchronous optical network (SONET) transport information model	Dec 1998
GR-1042-IMD	Generic requirements for operations interfaces using OSI tools—Information model detials: synchoronous optical network (SONET) transport information model	Dec 1998
GR-1230	SONET Bi-directional line-switched ring dquipment generic criteria	Dec 1998
GR-1244	Clocks for the Synchronized Network: Common Generic Criteria	Dec 2000
GR-1250	Generic requirements for synchornous optical network (SONET) file transfer	Dec 1999
GR-1332	Generic requirements for data communications network security	Apr 1996
GR-1345	Framework Generic Requirements for Element Manager (EM) Applications for SONET Subnetworks	Dec 2000
GR-1365	SONET private line service interface generic criteria for end users	Dec 1994
GR-1374	SONET inter-carrier interface physical layer generic criteria for carriers	Dec 1994
GR-1400	SONET dual-fed unidirectional path switched ring (UPSR) equipment generic criteria	Jan 1999

GR-1402	Network maintenance: Access and testing—DS3 HCDS TSC/RTU and DTAU functional requirements	Dec 1995
GR-253	Synchronous Optical Network (SONET) Transport Systems: Common Generic Criteria	Sep 2000
GR-2837	ATM virtual path ring functionality in SONET—generic criteria	Feb 1998
GR-2875	Generic requirement for digital interface systems	May 1996
GR-2891	SONET ATM virutal path digital cross-connect systems—generic criteria	Dec 1998
GR-2899	Generic criteria for SONET two-channel (1310/1550-NM) wavelength division multiplexed systems	Sep 1995
GR-2900	SONET asymmetric multiplex functional criteria	Sep 1995
GR-2950	Information Model for Sonet Digital Cross-Connect Systems (DCSS)	Feb 1999
GR-2955	Generic requirements for hybrid SONET/ATM element managements systems (EMSS)	Nov 1998
GR-2996	Generic criteria for SONET digital cross-connect systems	Jan 1999
GR-3000	Generic requirements for SONET element management systems	Nov 1999
GR-3001	Generic requirements for SONET network management systems (NMSs)	Dec 1999
GR-3003	Generic requirements for SONET/ATM network management systems	Feb 1999
GR-3004	Generic requirements for the operations interface between hybrid SONET/ATM element management systems and network management systems	Feb 1999

| GR-820 | OTGR section 5.1: generic digital transmission surveillance | Dec 1997 |
| GR-826 | OTGR section 10.2: User interface generic requirements for supporting network element operations | Jun 1994 |

ITU RECOMMENDATIONS

[G.801] Recommendation G.801 (11/88)   Digital transmission models.

[G.802] Recommendation G.802 (11/88)   Interworking between networks based on different digital hierarchies and speech encoding laws.

[G.803] Recommendation G.803 (03/00)   Architecture of transport networks based on the synchronous digital hierarchy (SDH). To be published.

[G.804] Recommendation G.804 (02/98)   ATM cell mapping into plesiochronous digital hierarchy (PDH).

Table of Contents of Recommendation G.804 (02/98)

[G.805] Recommendation G.805 (03/00)   Generic functional architecture of transport networks. To be published.

[G.806] Recommendation G.806 (10/00)   Characteristics of transport equipment. Description methodology and generic functionality. To be published.

[G.810] Recommendation G.810 (08/96)   Definitions and terminology for synchronization networks.

Summary of Recommendation G.810 (08/96)

[G.811] Recommendation G.811 (09/97)   Timing characteristics of primary reference clocks.

Table of Contents and Summary of Recommendation G.811 (09/97)

[G.812] Recommendation G.812 (06/98)   Timing requirements of slave clocks suitable for use as node clocks in synchronization networks.

Table of Contents and Summary of Recommendation G.812 (06/98)

[G.813] Recommendation G.813 (08/96)   Timing characteristics of SDH equipment slave clocks (SEC).

Summary of Recommendation G.813 (08/96)

[G.821] Recommendation G.821 (08/96)   Error performance of an international digital connection operating at a bit rate below the primary rate and forming part of an integrated services digital network.

Summary of Recommendation G.821 (08/96)

[G.822] Recommendation G.822 (11/88)   Controlled slip rate objectives on an international digital connection.

[G.823] Recommendation G.823 (03/00)   The control of jitter and wander within digital networks which are based on the 2048 kbit/s hierarchy. To be published.

[G.824] Recommendation G.824 (03/00)   The control of jitter and wander within digital networks which are based on the 1544 Kbps hierarchy.

[G.826] Recommendation G.826 (02/99)   Error performance parameters and objectives for international, constant bit rate digital paths at or above the primary rate.

Table of Contents and Summary of Recommendation G.826 (02/99)

[G.827] Recommendation G.827 (03/00)   Availability parameters and objectives for path elements of international constant bit-rate digital paths at or above the primary rate. To be published.

[G.827.1] Recommendation G.827.1 (11/00)   Availability performance objectives for end-to-end international constant bit-rate digital paths at or above the primary rate. To be published.

[G.828] Recommendation G.828 (03/00)   Error performance parameters and objectives for international, constant bit rate synchronous digital paths.

[G.831] Recommendation G.831 (03/00)   Management capabilities of transport networks based on the synchronous digital hierarchy (SDH). To be published.

[G.832] Recommendation G.832 (10/98)   Transport of SDH elements on PDH networks. Frame and multiplexing structures.

Table of Contents and Summary of Recommendation G.832 (10/98)

[G.841] Recommendation G.841 (10/98)   Types and characteristics of SDH network protection architectures.

Table of Contents and Summary of Recommendation G.841 (10/98)

[G.842] Recommendation G.842 (04/97)   Interworking of SDH network protection architectures.

Table of Contents of Recommendation G.842 (04/97)

[G.851.1] Recommendation G.851.1 (11/96)   Management of the transport network. Application of the RM-ODP framework.

[G.852.1] Recommendation G.852.1 (11/96)   Management of the transport network. Enterprise viewpoint for simple subnetwork connection management.

Summary of Recommendation G.852.1 (11/96)

[G.852.2] Recommendation G.852.2 (03/99)   Enterprise viewpoint description of transport network resource model.

Table of Contents of Recommendation G.852.2 (03/99)

[G.852.3] Recommendation G.852.3 (03/99)   Enterprise viewpoint for topology management.

Table of Contents and Summary of Recommendation G.852.3 (03/99)

[G.852.6] Recommendation G.852.6 (03/99)   Enterprise viewpoint for trail management.

Table of Contents and Summary of Recommendation G.852.6 (03/99)

[G.852.8] Recommendation G.852.8 (03/99)   Enterprise viewpoint for pre-provisioned adaptation management.

Table of Contents and Summary of Recommendation G.852.8 (03/99)

[G.852.10] Recommendation G.852.10 (03/99)   Enterprise viewpoint for pre-provisioned link connection management.

Table of Contents and Summary of Recommendation G.852.10 (03/99)

[G.852.12] Recommendation G.852.12 (03/99)   Enterprise viewpoint for pre-provisioned link management.

Table of Contents and Summary of Recommendation G.852.12 (03/99)

[G.852.16] Recommendation G.852.16 (01/01)   Enterprise viewpoint for pre-provisioned route discovery. To be published.

[G.853.1] Recommendation G.853.1 (03/99)   Common elements of the information viewpoint for the management of a transport network.

Table of Contents and Summary of Recommendation G.853.1 (03/99)

[G.853.2] Recommendation G.853.2 (11/96)   Subnetwork connection management information viewpoint.

Summary of Recommendation G.853.2 (11/96)

[G.853.3] Recommendation G.853.3 (03/99)   Information viewpoint for topology management.

Table of Contents and Summary of Recommendation G.853.3 (02/99)

[G.853.6] Recommendation G.853.6 (03/99)   Information viewpoint for trail management.

Table of Contents and Summary of Recommendation G.853.6 (03/99)

[G.853.8] Recommendation G.853.8 (03/99)   Information viewpoint for pre-provisioned adaptation management.

Table of Contents and Summary of Recommendation G.853.8 (03/99)

[G.853.10] Recommendation G.853.10 (03/99)   Information viewpoint for pre-provisioned link connection management.

Table of Contents and Summary of Recommendation G.853.10 (03/99)

[G.853.12] Recommendation G.853.12 (03/99)   Information viewpoint for pre-provisioned link management.

Table of Contents and Summary of Recommendation G.853.12 (03/99)

[G.853.16] Recommendation G.853.16 (01/01)   Information viewpoint for pre-provisioned route discovery. To be published.

[G.854.1] Recommendation G.854.1 (11/96)   Management of the transport network. Computational interfaces for basic transport network model.

Summary of Recommendation G.854.1 (11/96)

[G.854.3] Recommendation G.854.3 (03/99)   Computational viewpoint for topology management.

Table of Contents and Summary of Recommendation G.854.3 (03/99)

[G.854.6] Recommendation G.854.6 (03/99)   Computational viewpoint for trail management.

Table of Contents and Summary of Recommendation G.854.6 (03/99)

[G.854.8] Recommendation G.854.8 (03/99)   Computational viewpoint for pre-provisioned adaptation management.

Table of Contents and Summary of Recommendation G.854.8 (03/99)

[G.854.10] Recommendation G.854.10 (03/99)   Computational viewpoint for pre-provisioned link connection management.

Table of Contents and Summary of Recommendation G.854.10 (03/99)

[G.854.12] Recommendation G.854.12 (03/99)   Computational viewpoint for pre-provisioned link management.

Table of Contents and Summary of Recommendation G.854.12 (03/99)

[G.854.16] Recommendation G.854.16 (01/01)   Computational viewpoint for pre-provisioned route discovery. To be published.

[G.855.1] Recommendation G.855.1 (03/99)   GDMO engineering viewpoint for the generic network level model.

Table of Contents and Summary of Recommendation G.855.1 (03/99)

[G.861] Recommendation G.861 (08/96)   Principles and guidelines for the integration of satellite and radio systems in SDH transport networks.

Summary of Recommendation G.861 (08/96)

[G.871/Y.1301] Recommendation G.871/Y.1301 (10/00)   Framework for optical transport network Recommendations. To be published.

[G.872] Recommendation G.872 (02/99)   Architecture of optical transport networks.

Table of Contents and Summary of Recommendation G.872 (02/99)

[G.901] Recommendation G.901 (11/88)   General considerations on digital sections and digital line systems.

[G.902] Recommendation G.902 (11/95)   Framework Recommendation on functional access networks (AN) Architecture and functions, access types, management, and service node aspects.

Summary of Recommendation G.902 (11/95)

[G.911] Recommendation G.911 (04/97)   Parameters and calculation methodologies for reliability and availability of fiber optic systems.

Table of Contents and Summary of Recommendation G.911 (4/97)

[G.921] Recommendation G.921 (11/88)   Digital sections based on the 2048 kbit/s hierarchy.

[G.931] Recommendation G.931 (11/88)   Digital line sections at 3152 kbit/s.

[G.941] Recommendation G.941 (11/88)   Digital line systems provided by FDM transmission bearers.

[G.950] Recommendation G.950 (11/88)   General considerations on digital line systems.

[G.951] Recommendation G.951 (11/88)   Digital line systems based on the 1544 kbit/s hierarchy on symmetric pair cables.

[G.952] Recommendation G.952 (11/88)   Digital line systems based on the 2048 kbit/s hierarchy on symmetric pair cables.

[G.953] Recommendation G.953 (11/88)   Digital line systems based on the 1544 kbit/s hierarchy on coaxial pair cables.

[G.954] Recommendation G.954 (11/88)   Digital line systems based on the 2048 kbit/s hierarchy on coaxial pair cables.

[G.955] Recommendation G.955 (11/96)   Digital line systems based on the 1544 kbit/s and the 2048 kbit/s hierarchy on optical fiber cables.

Summary of Recommendation G.955 (11/96)

[G.957] Recommendation G.957 (06/99)   Optical interfaces for equipments and systems relating to the synchronous digital hierarchy.

Table of Contents and Summary of Recommendation G.957

[G.958] Recommendation G.958 (11/94)   Digital line systems based on the synchronous digital hierarchy for use on optical fiber cables.

[G.959.1] Recommendation G.959.1 (02/01) . Optical transport network physical layer interfaces. To be published.

[G.960] Recommendation G.960 (03/93) Access digital section for ISDN basic rate access.

[G.961] Recommendation G.961 (03/93) Digital transmission system on metallic local lines for ISDN basic rate access.

[G.961err] Covering Note to Recommendation G.961 (03/93)

[G.962] Recommendation G.962 (03/93) Access digital section for ISDN primary rate at 2048 kbit/s.

[G.962 Amend.1] Amendment 1 (06/97) to Recommendation G.962 Access digital section for ISDN primary rate at 2048 kbit/s. Amendment 1: Maintenance channel.

Table of Contents of Amendment 1 (06/97) to Recommendation G.962

[G.963] Recommendation G.963 (03/93) Access digital section for ISDN primary rate at 1544 kbit/s.

[G.964] Recommendation G.964 (03/01) V-Interfaces at the digital local exchange (LE). V5.1-Interface (based on 2048 kbit/S) for the support of access network (AN). To be published.

[G.965] Recommendation G.965 (03/01) V-Interfaces at the digital local exchange (LE). V5.2 interface (based on 2048 kbit/s) for the support of access network (AN). To be published.

[G.966] Recommendation G.966 (02/99) Access digital section for B-ISDN.

Table of Contents and Summary of Recommendation G.966 (02/99)

[G.967.1] Recommendation G.967.1 (06/98) V-interfaces at the service node (SN): VB5.1 reference point specification.

Table of Contents and Summary of Recommendation G.967.1 (06/98)

[G.967.2] Recommendation G.967.2 (02/99) V-interfaces at the service node (SN): VB5.2 reference point specification.

Table of Contents and Summary of Recommendation G.967.2 (02/99)

[G.967.3] Recommendation G.967.3 (03/00) V-interfaces at the service node (SN): Protocol implementation conformance statements for interfaces at VB5 reference points.

[G.971] Recommendation G.971 (04/00) General features of optical fibre submarine cable systems. To be published.

[G.972] Recommendation G.972 (10/00) Definition of terms relevant to optical fiber submarine cable systems. To be published.

[G.973] Recommendation G.973 (11/96) Characteristics of repeaterless optical fiber submarine cable systems.

Summary of Recommendation G.973 (11/96)

[G.974] Recommendation G.974 (03/93) Characteristics of regenerative optical fiber submarine cable systems.

[G.975] Recommendation G.975 (10/00) Forward error correction for submarine systems. To be published.

[G.976] Recommendation G.976 (10/00) Test methods applicable to optical fiber submarine cable systems. To be published.

[G.977] Recommendation G.977 (04/00) Characteristics of optically amplified optical submarine cable systems. To be published.

[G.981] Recommendation G.981 (01/94) PDH optical line systems for the local network.

[G.982] Recommendation G.982 (11/96) Optical access networks to support services up to the ISDN primary rate or equivalent bit rates.

Summary of Recommendation G.982 (11/96)

[G.983.1] Recommendation G.983.1 (10/98) Broadband optical access systems based on Passive Optical Networks (PON).

Table of Contents and Summary of Recommendation G.983.1 (10/98)

[G.983.1 Corr.1] Corrigendum 1 (06/99) to Recommendation G.983.1 Broadband optical access systems based on Passive Optical Network (PON).

[G.983.2] Recommendation G.983.2 (04/00) ONT management and control interface specification for ATM PON. To be published.

[G.983.3] Recommendation G.983.3 (02/01) A broadband optical access system with increased service capability by wavelength allocation. To be published.

[G.989.1] Recommendation G.989.1 (02/01) Phoneline networking transceivers. Foundation. To be published.

[G.991.1] Recommendation G.991.1 (10/98) High bit rate Digital Subscriber Line (HDSL) transceivers.

Table of Contents and Summary of Recommendation G.991.1 (10/98)

[G.991.2] Recommendation G.991.2 (02/01) Single-pair high-speed digital subscriber line (SHDSL) transceivers. For approval. Updated. To be published.

[G.992.1] Recommendation G.992.1 (06/99) Asymmetric Digital Subscriber Line (ADSL) transceivers.

[G.992.1 Ann.H] Annex H to Recommendation G.992.1 (10/00) Specific requirements for a synchronized symmetrical DSL (SSDSL) system operating in the same cable binder as ISDN as defined in ITU-T Recommendation G.961 Appendix III. To be published.

[G.992.2] Recommendation G.992.2 (06/99) Splitterless asymmetric digital subscriber line (ADSL) transceivers.

Table of Contents and Summary of Recommendation G.992.2 (06/99)

[G.994.1] Recommendation G.994.1 (02/01) Handshake procedures for Digital Subscriber Line (DSL) transceivers. To be published.

[G.995.1] Recommendation G.995.1 (02/01) Overview of digital subscriber line (DSL) recommendations. For approval. To be published.

[G.996.1] Recommendation G.996.1 (02/01) Test procedures for digital subscriber line (DSL) transceivers. To be published.

[G.997.1] Recommendation G.997.1 (06/99) Physical layer management for digital subscriber line (DSL) transceivers.

Table of Contents and Summary of Recommendation G.997.1 (06/99)

[Supp-37] Supplement 37 to ITU-T Series G Recommendations ITU-T Recommendation G.763 digital circuit multiplication equipment (DCME) tutorial and dimensioning.

[Supp-38] Supplement 38 to ITU-T Series G Recommendations Variable bit rate calculations for ITU-T Recommendation G.767 Digital Circuit Multiplication Equipment (DCME).

[G.701] Recommendation G.701 (03/93) Vocabulary of digital transmission and multiplexing, and pulse code modulation (PCM) terms.

[G.702] Recommendation G.702 (11/88) Digital hierarchy bit rates.

[G.703] Recommendation G.703 (10/98) Physical/electrical characteristics of hierarchical digital interfaces.

Table of Contents and Summary of Recommendation G.703 (10/98)

[G.704] Recommendation G.704 (10/98) Synchronous frame structures used at 1544, 6312, 2048, 8448, and 44 736 kbit/s hierarchical levels.

Table of Contents and Summary of Recommendation G.704 (10/98)

[G.705] Recommendation G.705 (10/00) Characteristics of Plesiochronous Digital Hierarchy (PDH) equipment functional blocks. To be published.

[G.706] Recommendation G.706 (04/91) Frame alignment and cyclic redundancy check (CRC) procedures relating to basic frame structures defined in Recommendation G.704.

Summary of Recommendation G.706 (04/91)

[G.707/Y.1322] Recommendation G.707/Y.1322 (10/00) Network node interface for the synchronous digital hierarchy (SDH). To be published.

[G.707 Corr.1] Corrigendum 1 (03/01) to Recommendation [G.707/Y.1322.] To be published.

[G.708] Recommendation G.708 (06/99) Sub STM-0 network node interface for the synchronous digital hierarchy (SDH).

Table of Contents and Summary of Recommendation G.708 (06/99)

[G.709] Recommendation G.709 (02/01) Interface for the optical transport network (OTN). To be published.

[G.711] Recommendation G.711 (11/88) Pulse code modulation (PCM) of voice frequencies.

[G.711 App.I] Appendix I (09/99) to Recommendation G.711 A high quality low-complexity algorithm for packet loss concealment with G.711.

[G.711 App.II] Appendix II (02/00) to Recommendation G.711 A comfort noise payload definition for ITU-T G.711 use in packet-based multimedia communication systems.

[G.712] Recommendation G.712 (11/96) Transmission performance characteristics of pulse code modulation channels.

Summary of Recommendation G.712 (11/96)

[G.720] Recommendation G.720 (07/95) Characterization of low-rate digital voice coder performance with non-voice signals.

[G.722] Recommendation G.722 (11/88) 7 kHz audio-coding within 64 kbit/s.

[G.722 Annex A] Annex A to Recommendation G.722 (03/93) Testing signal-to-total distortion ratio for 7 kHz audio-codecs at 64 kbit/s. Recommendation G.722 connected back-to-back.

[G.722 App.II] Appendix II to Recommendation G.722 (03/87) Description of the digital test sequences for the verification of the G.722 64 kbit/s SB-ADPCM 7 kHz coded.

[G.722.1] Recommendation G.722.1 (09/99)   Coding at 24 and 32 kbit/s for hands-free operation in systems with low frame loss. To be published.

[G.722.1Annex A] Annex A (02/00) to Recommendation G.722.1 Packet format, capability identifiers, and capability parameters.

Table of Contents of Annex A to Recommendation G.722.1 (02/00)

[G.722.1Annex B] Annex B (11/00) to Recommendation G.722.1 Floating point implementation for G.722.1. To be published.

[G.722.1 Corr.1] Corrigendum 1 (11/00) to Recommendation G.722.1 To be published.

[G.723.1] Recommendation G.723.1 (03/96)   Dual rate speech coder for multimedia communications transmitting at 5.3 and 6.3 kbit/s.

Summary of Recommendation G.723.1 (03/96)

[G.723.1Annex A] Annex A (11/96) to Recommendation G.723.1 Dual rate speech coder for multimedia communications transmitting at 5.3 and 6.3 kbit/s. Silence compression scheme.

Table of Contents of Annex A to Recommendation G.723.1

[G.723.1Annex B] Annex B (11/96) to Recommendation G.723.1 Dual rate speech coder for multimedia communications transmitting at 5.3 and 6.3 kbit/s. Annex B: Alternative specification based on floating point arithmetic.

Table of Contents of Annex B to Recommendation G.723.1

[G.723.1Annex C] Annex C (11/96) to Recommendation G.723.1 Dual rate speech coder for multimedia communications transmitting at 5.3 and 6.3 kbit/s. Annex C: Scalable channel coding scheme for wireless applications.

Table of Contents of Annex C to Recommendation G.723.1

[G.724] Recommendation G.724 (11/88) Characteristics of a 48-channel low bit rate encoding primary multiplex operating at 1544 kbit/s.

[G.725] Recommendation G.725 (11/88)   System aspects for the use of the 7 kHz audio codec within 64 kbit/s.

[G.726] Recommendation G.726 (12/90)   40, 32, 24, and 16 kbit/s Adaptive Differential Pulse Code Modulation (ADPCM).

Summary of Recommendation G.726 (12/90)

[G.726 Annex A] Annex A to Recommendation G.726 (11/94)   Extensions of Recommendation G.726 for use with uniform-quantized input and output.

[G.726 App. II] Appendix II Test Vectors to Recommendation G.726 (03/91)   Description of the digital test sequences for the verification of the G.726 40, 32, 24, and 16 kbit/s ADPCM algorithm.

[AppG.726/G.727] Appendix III (Rec. G.726)/Appendix II (Rec. G.727) (05/94)   Comparison of ADPCM algorithms.

[G.727] Recommendation G.727 (12/90)   5-, 4-, 3-, and 2-bits per sample embedded adaptive differential pulse code modulation (ADPCM).

Summary of Recommendation G.727 (12/90)

[G.727 Annex A] Annex A to Recommendation G.727 (11/94)   Extensions of Recommendation G.727 for use with uniform-quantized input and output.

[G.727 App. I] Appendix I to Recommendation G.727 (03/91)   Description of the digital test sequences for the verification of the G.727 5-, 4-, 3-, and 2-bits per sample embedded ADPCM algorithm.

[G.728] Recommendation G.728 (09/92)   Coding of speech at 16 kbit/s using low-delay code excited linear prediction.

[G.728 App. 1] Appendix I (07/95) to Recommendation G.728   Verification tools. Programs and test sequences for implementation verification of the algorithm of the G.728 16 kbit/s LD-CELP speech coder.

[G.728 App. II] Appendix II (11/95) to Recommendation G.728   Speech performance.

[G.728 Annex G] Annex G (11/94) to Recommendation G.728   Coding of speech at 16 kbit/s using low-delay code excited linear prediction. Annex G: 16 kbit/s fixed point specification.

[G.728Cor.1An.G] Corrigendum 1 (02/00) of Annex G to Recommendation G.728   16 kbit/s fixed point specification.

[G.728 Annex H] Annex H (05/99) to Recommendation G.728   Variable bit rate LD-CELP operation mainly for DCME at rates less than 16 kbit/s.

Table of Contents of Annex H to Recommendation G.728

[G.728 Annex I] Annex I (05/99) to Recommendation G.728   Frame or packet loss concealment for the LD-CELP decoder.

Table of Contents and Summary of Annex I to Recommendation G.728 (05/99)

[G.728 Annex J] Annex J (09/99) to Recommendation G.728   Variable bit-rate operation of LD-CELP mainly for voiceband-data applications in DCME.

Table of Contents and Summary of Annex J to Recommendation G.728 (09/99)

[G.729] Recommendation G.729 (03/96) C source code and test vectors for implementation verification of the G.729 8 kbit/s CS-ACELP speech coder.

Table of Contents of Recommendation G.729 (03/96)

[G.729 Annex A] Annex A (11/96) to Recommendation G.729 C source code and test vectors for implementation verification of the G.729 reduced complexity 8 kbit/s CS-ACELP speech coder.

Table of Contents of Annex A to Recommendation G.729 (03/96)

[G.729 Annex B] Annex B (10/96) to Recommendation G.729 A silence compression scheme for G.729 optimized for terminals conforming to Recommendation V.70.

Table of Contents and Summary of Annex B to Recommendation G.729 (10/96)

[G.729 Annex C] Annex C (09/98) to Recommendation G.729 Coding of speech at 8 kbit/s using Conjugate-Structure Algebraic-Code-Excited-Linear-Prediction (CS-ACELP). Annex C: Reference floating-point implementation for G.729 CS-ACELP 8 kbit/s speech coding.

Table of Contents and Summary of Annex C to Recommendation G.729 (09/98)

[G.729 Annex C+] Annex C+ (02/00) to Recommendation G.729 Annex C+: Reference floating-point implementation for integrating G.729 CS-ACELP speech coding main body with Annexes B, D, and E.

Table of Contents and Summary of Annex C+ to Recommendation G.729 (02/00)

[G.729 Annex D] Annex D (09/98) to Recommendation G.729 Coding of speech at 8 kbit/s using Conjugate Structure Algebraic-Code-Excited Linear-Prediction (CS-ACELP). Annex D: 6.4 kbit/s CS-ACELP speech coding algorithm.

Table of Contents and Summary of Annex D to Recommendation G.729 (02/00)

[G.729 Annex E] Annex E (09/98) to Recommendation G.729 Coding of speech at 8 kbit/s using Conjugate-Structure Algebraic-Code-Excited-Linear-Prediction (CS-ACELP). Annex E: 11.8 kbit/s CS-ACELP speech coding algorithm.

Table of Contents and Summary of Annex E to Recommendation G.729 (09/98)

[G.729 Annex F] Annex F (02/00) to Recommendation G.729 DTX functionality for G.729 Annex D.

Table of Contents and Summary of Annex F to Recommendation G.729 (02/00)

[G.729 Annex G] Annex G (02/00) to Recommendation G.729 DTX functionality for G.729 Annex E.

Table of Contents and Summary of Annex G to Recommendation G.729 (02/00)

[G.729 Annex H] Annex H (02/00) to Recommendation G.729 Reference implementation of switching procedure between G.729 Annexes D and E.

Table of Contents and Summary of Annex H to Recommendation G.729 (02/00)

[G.729 Annex I] Annex I (02/00) to Recommendation G.729 Integrated C code for G.729 main body, Annexes B, D, and E.

Table of Contents and Summary of Annex I to Recommendation G.729 (02/00)

[G.729CorrAnnex] Corrigendum (03/01) to Annexes of Recommendation G.729. To be published.

[G.731] Recommendation G.731 (11/88) Primary PCM multiplex equipment for voice frequencies.

[G.732] Recommendation G.732 (11/88) Characteristics of primary PCM multiplex equipment operating at 2048 kbit/s.

[G.733] Recommendation G.733 (11/88) Characteristics of primary PCM multiplex equipment operating at 1544 kbit/s.

[G.734] Recommendation G.734 (11/88) Characteristics of synchronous digital multiplex equipment operating at 1544 kbit/s.

[G.735] Recommendation G.735 (11/88) Characteristics of primary PCM multiplex equipment operating at 2048 kbit/s and offering synchronous digital access at 384 kbit/s and/or 64 kbit/s.

[G.736] Recommendation G.736 (03/93) Characteristics of a synchronous digital multiplex equipment operating at 2048 kbit/s.

[G.737] Recommendation G.737 (11/88) Characteristics of an external access equipment operating at 2048 kbit/s offering synchronous digital access at 384 kbit/s and/or 64 kbit/s.

[G.738] Recommendation G.738 (11/88) Characteristics of primary PCM multiplex equipment operating at 2048 kbit/s and offering synchronous digital access at 320 kbit/s and/or 64 kbit/s.

[G.739] Recommendation G.739 (11/88) Characteristics of an external access equipment operating at 2048 kbit/s offering synchronous digital access at 320 kbit/s and/or 64 kbit/s.

[G.741] Recommendation G.741 (11/88) General considerations on second order multiplex equipments.

[G.742] Recommendation G.742 (11/88) Second order digital multiplex equipment operating at 8448 kbit/s and using positive justification.

[G.743] Recommendation G.743 (11/88) Second order digital multiplex equipment operating at 6312 kbit/s and using positive justification.

[G.744] Recommendation G.744 (11/88) Second order PCM multiplex equipment operating at 8448 kbit/s.

[G.745] Recommendation G.745 (11/88) Second order digital multiplex equipment operating at 8448 kbit/s and using positive/zero/negative justification.

[G.746] Recommendation G.746 (11/88) Characteristics of second order PCM multiplex equipment operating at 6312 kbit/s.

[G.747] Recommendation G.747 (11/88) Second order digital multiplex equipment operating at 6312 kbit/s and multiplexing three tributaries at 2048 kbit/s.

[G.751] Recommendation G.751 (11/88) Digital multiplex equipments operating at the third order bit rate of 34 368 kbit/s and the fourth order bit rate of 139 264 kbit/s and using positive justification.

[G.752] Recommendation G.752 (11/88) Characteristics of digital multiplex equipments based on a second order bit rate of 6312 kbit/s and using positive justification.

[G.753] Recommendation G.753 (11/88) Third order digital multiplex equipment operating at 34 368 kbit/s and using positive/zero/negative justification.

[G.754] Recommendation G.754 (11/88) Fourth order digital multiplex equipment operating at 139 264 kbit/s and using positive/zero/negative justification.

[G.755] Recommendation G.755 (11/88) Digital multiplex equipment operating at 139 264 kbit/s and multiplexing three tributaries at 44 736 kbit/s.

[G.761] Recommendation G.761 (11/88) General characteristics of a 60-channel transcoder equipment.

[G.762] Recommendation G.762 (11/88) General characteristics of a 48-channel transcoder equipment.

[G.763] Recommendation G.763 (10/98) Digital circuit multiplication equipment using G.726 ADPCM and digital speech interpolation.

Table of Contents and Summary of Recommendation G.763 (10/98)

[G.763err] Covering Note to Recommendation G.763 (10/98)

[G.764] Recommendation G.764 (12/90) Voice packetization. Packetized voice protocols.

Summary of Recommendation G.764 (12/90)

[G.764 App. I] Appendix I (11/95) to Recommendation G.764 Packetization guide.

Summary of Appendix I (11/95) to Recommendation G.764

[G.765] Recommendation G.765 (09/92) Packet circuit multiplication equipment.

[G.765 App. I] Appendix I (11/95) to Recommendation G.765 A guide to PCME.

[G.766] Recommendation G.766 (11/96) Facsimile demodulation/ remodulation for digital circuit multiplication equipment.

[G.767] Recommendation G.767 (10/98) Digital circuit multiplication equipment using 16 kbit/s LD-CELP, digital speech interpolation, and facsimile demodulation/remodulation.

Table of Contents and Summary of Recommendation G.767 (10/98)

[G.768] Recommendation G.768 (03/01) Digital circuit multiplication equipment using 8 kbit/s CS-ACELP. To be published.

[G.772] Recommendation G.772 (03/93) Protected monitoring points provided on digital transmission systems.

[G.773] Recommendation G.773 (03/93) Protocol suites for Q-interfaces for management of transmission systems.

[G.774] Recommendation G.774 (02/01) Synchronous digital hierarchy (SDH) management information model for the network element view. To be published.

[G.774.1] Recommendation G.774.1 (02/01) Synchronous digital hierarchy (SDH) bidirectional performance monitoring for the network element view. To be published.

[G.774.2] Recommendation G.774.2 (02/01) Synchronous digital hierarchy (SDH) configuration of the payload structure for the network element view. To be published.

[G.774.3] Recommendation G.774.3 (02/01) Synchronous digital hierarchy (SDH) management of multiplex-section protection for the network element view. To be published.

[G.774.4] Recommendation G.774.4 (02/01) Synchronous digital hierarchy (SDH) management of the subnetwork connection protection for the network element view. To be published.

[G.774.5] Recommendation G.774.5 (02/01) Synchronous digital hierarchy (SDH) management of connection supervision functionality (HCS/LCS) for the network element view. To be published.

[G.774.6] Recommendation G.774.6 (02/01) Synchronous digital hierarchy (SDH) unidirectional performance monitoring for the network element view. To be published.

[G.774.7] Recommendation G.774.7 (02/01) Synchronous digital hierarchy (SDH) management of lower order path trace and interface labelling for the network element view. To be published.

[G.774.8] Recommendation G.774.8 (02/01) Synchronous digital hierarchy (SDH) management of radio-relay systems for the network element view. To be published.

[G.774.9] Recommendation G.774.9 (02/01) Synchronous Digital Hierarchy (SDH) configuration of linear multiplex section protection for the network element view. To be published.

[G.774.10] Recommendation G.774.10 (02/01) Synchronous Digital Hierarchy (SDH) Multiplex Section (MS) shared protection ring management for the network element view. To be published.

[G.775] Recommendation G.775 (10/98) Loss of Signal (LOS), Alarm Indication Signal (AIS), and Remote Defect Indication (RDI) defect detection and clearance criteria for PDH signals.

Table of Contents and Summary of Recommendation G.775 (10/98)

[G.776.1] Recommendation G.776.1 (10/98) Managed objects for signal processing network elements.

Table of Contents and Summary of Recommendation G.776.1 (10/98)

[G.776.3] Recommendation G.776.3 (04/00) ADPCM DCME Configuration Map Report.

[G.780] Recommendation G.780 (06/99) Vocabulary of terms for synchronous digital hierarchy (SDH) networks and equipment.

Table of Contents and Summary of Recommendation G.780 (06/99)

[G.781] Recommendation G.781 (06/99) Synchronization layer functions.

Table of Contents and Summary of Recommendation G.781

[G.783] Recommendation G.783 (10/00) Characteristics of synchronous digital hierarchy (SDH) equipment functional blocks. To be published.

[G.783 Corr.1] Corrigendum 1 (03/01) to Recommendation G.783 To be published.

[G.784] Recommendation G.784 (06/99) Synchronous digital hierarchy (SDH) management.

Table of Contents and Summary of Recommendation G.784 (06/99)

[G.785] Recommendation G.785 (11/96) Characteristics of a flexible multiplexer in a synchronous digital hierarchy environment.

Summary of Recommendation G.785 (11/96)

[G.791] Recommendation G.791 (11/88) General considerations on transmultiplexing equipments.

[G.792] Recommendation G.792 (11/88) Characteristics common to all transmultiplexing equipments.

[G.793] Recommendation G.793 (11/88) Characteristics of 60-channel transmultiplexing equipments.

[G.794] Recommendation G.794 (11/88) Characteristics of 24-channel transmultiplexing equipments.

[G.795] Recommendation G.795 (11/88) Characteristics of codecs for FDM assemblies.

[G.796] Recommendation G.796 (09/92) Characteristics of a 64 kbit/s cross-connect equipment with 2048 kbit/s access ports.

[G.796 Corr.1] Corrigendum 1 (10/98) to Recommendation G.796 Characteristics of a 64 kbit/s cross-connect equipment with 2048 kbit/s access ports Corrigendum 1.

[G.797] Recommendation G.797 (03/96) Characteristics of a flexible multiplexer in a plesiochronous digital hierarchy environment.

ANSI DOCUMENTS

T1.105.01–2000

Synchronous Optical Network (SONET)—Automatic Protection

ANSI T1.105.09–1996

Synchronous Optical Network (SONET)—Network Element Timing and Synchronization

ANSI T1.105.07a–1997

Supplement ANSI T1.105.07a–1997

ANSI T1.105.07–1996

Synchronous Optical Network (SONET)—Sub-STS-1 Interface Rates and Formats Specification

ANSI T1.105.06–1996

Synchronous Optical Network (SONET)—Physical Layer Specification (Revision of ANSI T1.106–1988)

ANSI T1.105.05–1994

Synchronous Optical Network (SONET)—Tandem Connection Maintenance

ANSI T1.105.04–1995

Synchronous Optical Network (SONET)—Data Communication Channel Protocol and Architectures

ANSI T1.105.03b–1997

Supplement ANSI T1.105.03b–1997

ANSI T1.105.03a–1995

Supplement ANSI T1.105.03a–1995

ANSI T1.105.03–1994

Synchronous Optical Network (SONET)—Jitter at Network Interfaces

ANSI T1.105.02–1995

Synchronous Optical Network (SONET)—Payload Mappings

ANSI T1.105–1995

Synchronous Optical Network (SONET)—Basic Description including Multiplex Structure, Rates, and Formats

PHOTO CREDITS

All photographs and illustrations are from the author, except where noted below:

Figures 1-11, 1-12, 1-13 and 4-40 are used courtesy of Lucent Technologies.
Figure 2-17 is used courtesy of the Port of Oakland, Oakland, California.
Figures 4-17 and 4-18 are used courtesy of Todd Quam.

INDEX

O

R

S

U

ABOUT THE AUTHOR

Steve Shepard is a professional writer and educator, specializing in international telecommunications. He teaches technical courses in corporations throughout the world. Formerly with Hill Associates, he is the author of *Telecommunications Convergence* and *Optical Networking Demystified*, also from McGraw-Hill. He is based in Williston, Vermont.